ORGANIC SOILS
and
PEAT MATERIALS
for
SUSTAINABLE
AGRICULTURE

ORGANIC SOILS
and
PEAT MATERIALS
for
SUSTAINABLE
AGRICULTURE

Edited by
Léon-Etienne Parent
Piotr Ilnicki

CRC Press
Taylor & Francis Group
Boca Raton London New York

CRC Press is an imprint of the
Taylor & Francis Group, an **informa** business

Cover photograph courtesy of Léon-Etienne Parent

Published 2003 by CRC Press
Taylor & Francis Group
6000 Broken Sound Parkway NW, Suite 300
Boca Raton, FL 33487-2742

© 2003 by Taylor & Francis Group, LLC
CRC Press is an imprint of Taylor & Francis Group, an Informa business

First issued in paperback 2019

No claim to original U.S. Government works

ISBN-13: 978-0-367-45484-5 (pbk)
ISBN-13: 978-0-8493-1458-2 (hbk)

Visit the Taylor & Francis Web site at
http://www.taylorandfrancis.com

and the CRC Press Web site at
http://www.crcpress.com

Library of Congress Cataloging-in-Publication Data

Organic soils and peat materials for sustainable agriculture / edited by Léon-Etienne Parent, Piotr Ilnicki.
    p. cm.
  Includes bibliographical references and index.
  ISBN 0-8493-1458-5 (alk. paper)
    1. Histosols. 2. Peat soils. 3. Sustainable agriculture. I. Parent, Léon-Etienne. II. Ilnicki, Piotr.

S598 .O725 2002
631.4'17—dc21   2002022951

Library of Congress Card Number 2002022951

Dedication

To the memory of Professor Henryk Okruszko (1925–2000),
Research Soil Scientist, Institute of Land Reclamation and Grassland
Farming (IMUZ) at Falenty, Poland, who described and documented
the moorsh-forming process as a generic soil concept for the
management of drained organic soils.

Auspices Declaration

The International Peat Society, founded in Québec, Canada, in 1968, is a nongovernmental, nonprofit organization. Its mission is to promote wise use of mires, peatlands, and peat by advancing scientific, technical, economic, and social knowledge and understanding.

The members are involved in the study and use of peat, for instance, in agriculture, forestry, horticulture, energy, chemical technology, environmental protection, balneology, medicine, and related areas. Therefore, they explore a broad range of peat-related issues.

Research institutes, commercial companies, government, nongovernment organizations, as well as private individuals form a unique base not only for information exchange concerning the role of peatlands in the world's geology, hydrology, climate, and ecology, but also for their wise and sustainable utilization.

To promote knowledge and cooperation among its members, the International Peat Society organizes international congresses, symposia, and workshops and provides several publications, such as the *International Peat Journal*. The Secretariat of the International Peat Society is located in Jyväskylä, Finland.

International Peat Society
Jyväskylä, Vapaudenkatu 1240100
Finland
Tel. +358 14 3385 440
Fax. + 358 14 3385 410
ips@peatsociety.fi

Foreword

The Ramsar Convention defined the wise use of peatlands as "their sustainable use for the benefit of mankind in a way compatible with the maintenance of the natural properties of the ecosystem." Land sustainability relates to definite periods and land uses. According to the International Peat Society (IPS) and the International Mire Conservation Group (IMCG), the wise use of peatlands in forestry and agriculture should include careful land-use planning and water management sustaining socioeconomic development in rural areas and preservation of flora and fauna (Peatlands International 2/1998: 25). Therefore, peatland management must account for unacceptable ecological impacts on the contiguous and global environment. It is well recognized that unsuitable practices must be alleviated within a relatively short timeframe (i.e., a few years). This book, initiated during the IPS Congress held in Bremen in 1996 by an international working group of IPS Commission III, is an attempt to define organic soil quality attributes leading to a wiser soil management in peatlands used for agriculture.

Irreversible processes occurring in drained organic soils must be managed cautiously. The moorsh-forming process is the main step in the degradation of organic soil materials, driving the physical deformation of peat colloids, peat decomposition and related processes such as CO_2 production, organic N and P mineralization, nitrification and denitrification, as well as the leaching of inorganic phosphate, nitrate, and soluble carbon. Those processes are at risk to the contiguous and global environment, and must be constrained during the short-living exploitation of organic soils. There are many scientific and technical reports indicating significant contribution of drained organic soils to off-farm water pollution by nitrate, phosphate and soluble C, and to greenhouse gas emissions, primarily as CO_2 and N_2O. On the other hand, organic soils contribute greatly to the production of food and ornamental plants to the benefit of mankind. The authors hope that this book will prompt action in the field and stimulate research for the wise management of organic soils and peat materials, shifting the present paradigm of input-based and unsustainable use to the modern paradigm of knowledge-based and sustainable agriculture and horticulture.

The authors collected up-to-date information in order to define quality indicators for the optimum use of organic soils in the agricultural and horticultural productions. The soil quality attributes make up its pedigree (from the French expression *pied-de-grue* or crane foot, meaning the three typical traits of a thing — living or nonliving, similarly to a crane footprint). The soil *pied-de-grue* is documented through its physical, chemical and biological attributes. Soil quality can be assessed both quantitatively or qualitatively. Soil quality is related to sustainable productivity, and to its limited capacity to act as an environmental buffer. Soil buffering capacity, as related to organic matter content and pH, comprises resistance to carbon, nitrogen, and phosphorus release, as well as water, pesticide, and metal retention. Soil conservation practices include water regulation, crop cover, fertilization, and liming, as well as soil restoration after peat wastage or cutting.

This book is composed of 10 chapters: the characteristics of organic soil genesis and degradation, (Chapter 1), the irreversible peat drying in drained organic soils

(Chapter 2), the physical attributes required for the water management of organic soils (Chapter 3) and peat substrates (Chapter 4), pH determination and correlation among methods (Chapter 5), the N and P pollution indicators (Chapter 6), Cu sorption (Chapter 7), pesticide reactions (Chapter 8), cultivation of cutover peatlands (Chapter 9), and conservation of organic soils (Chapter 10). The book covers a large spectrum of organic soil materials used *in situ* or in containers.

Léon E. Parent, chairman of the IPS-WG
Department of Soil Science and Agri-Food Engineering
Laval University, Québec, Canada G1K 7P4

The Editors

Dr. Léon Etienne Parent received his Ph.D. in soil fertility from McGill University, Montreal, Canada. He is a member of the Soil Science Society of America, the Canadian Soil Science Society, the International Soil Science Society, and the International Peat Society (IPS). He has chaired the Histic Soil Quality Working Group of IPS since 1996. He has authored or coauthored over 60 peer-reviewed articles, more than half related to organic soils and peat substrates. He has also contributed to seven book chapters and technical papers on organic soils. His major contributions to organic soil science concern soil transformations after drainage, agricultural capability classification, crop rotation and nutrient diagnosis, as well as the use of peat as a mineral soil amendment and as a source of humic substances for enhancing phosphorus fertilizer efficiency in mineral soils.

Dr. Piotr Ilnicki received his Ph.D. in soil science from the Agricultural University in Szczecin, Poland. He is a former vice-president of the International Peat Society (IPS), and president of Commission III of IPS (agriculture) and of the Polish National Committee of IPS. He is member of the Polish Soil Science Society and of the International Association for Landscape Ecology. He has authored or coauthored more than 70 peer-reviewed articles. He also edited a Polish handbook on peatlands and peat.

Contributors

Tomasz Brandyk
Department of Environmental
  Development and Land Improvement
Warsaw Agricultural University
Warsaw, Poland

Jean Caron
Department of Soil Science and
  Agri-Food Engineering
Laval University
Québec, Canada

Caroline Côté
Institut de Recherche et de
  Développement en Agro-
  Environnement (IRDA)
Saint-Hyacinthe, Québec, Canada

Josée Fortin
Department of Soil Science and Agri-
  Food Engineering
Laval University
Québec, Canada

Tomasz Gnatowski
Department of Environmental
  Development and Land Improvement
Warsaw Agricultural University
Warsaw, Poland

Piotr Ilnicki
Department of Environmental
  Protection and Management
Agricultural University
Poznan, Poland

Antoine Karam
Department of Soil Science and
  Agri-Food Engineering
Laval University
Québec, Canada

Lotfi Khiari
Department of Soil Science and
  Agri-Food Engineering
Laval University
Québec, Canada

Vera N. Kreshtapova
Dokutchaev Soil Science Institute
Moscow, Russia

Rudolf A. Krupnov
Tver State Technical University
Tver, Russia

Henryk Okruszko
Institute of Land Reclamation and
  Grassland Farming
Falenty, Poland

Ryszard Oleszczuk
Department of Environmental
  Development and Land Improvement
Warsaw Agricultural University
Warsaw, Poland

Léon E. Parent
Department of Soil Science and
  Agri-Food Engineering
Laval University
Québec, Canada

Louis-Marie Rivière
Unité Mixte de Recherches
Sciences Agronomiques Appliquées à
  l'Horticulture
Institut National de la Recherche
  Agronomique
Institut National d'Horticulture
Beaucouzé, France

Jan Szatylowicz
Department of Environmental
  Development and Land Improvement
Warsaw Agricultural University
Warsaw, Poland

Catherine Tremblay
Department of Soil Science and
  Agri-Food Engineering
Laval University
Québec, Canada

Olga N. Uspenskaya
Dokutchaev Soil Science Institute
Moscow, Russia

Jutta Zeitz
Institute of Crop Sciences
Faculty of Agriculture and Horticulture
Humboldt University
Berlin, Germany

Contents

CHAPTER 1

The Moorsh Horizons as Quality Indicators of Reclaimed Organic Soils

Henryk Okruszko and Piotr Ilnicki

CONTENTS

ABSTRACT

In Poland, drained mires are called peatlands. After reclamation, organic soils, primarily those originating from fens undergo genetical transformations designated globally as the moorsh-forming process (MFP). The intensity of MFP is monitored by changes in soil morphology and structure such as peat transforming into grains and aggregates, thus altering physical properties. The descriptive criteria for those transformations are as follows:

- Three diagnostic horizons (moorsh layers M_1 to M_3 in the 0–30 cm root zone, peat layers T_1 in the 30–80 cm vadose zone, and peat layers T_2 in the 80–130 cm zone of lateral groundwater flow)
- Three genetic layers in the vertical sequence M_1 to M_3 (M_1 as grain moorsh in the sod, M_2 as humic moorsh, and M_3 as peaty moorsh above peat layers)

- Three stages of MFP development (I, II, and III) in the moorsh layers; and three degrees of peat decomposition (a for fibric, b for hemic, and c for sapric materials) for T_1 and T_2

For water management, five prognostic soil moisture complexes are recognized from those criteria: wet (A), periodically wet (AB), moist (B), periodically dry (BC), and dry (C). Irrigation and water table regulation must be carefully designed for improving soil quality in BC and C complexes.

I. INTRODUCTION

In Poland, a wetland showing peat accumulation is designated as a mire for undrained, or as a peatland for drained, conditions. Mires are further classified according to trophic levels into fens, transitional bogs, and raised bogs. Organic soils are concerned with two major soil processes: paludification for peat accumulating under waterlogged conditions, and decession with moorsh forming and peat mineralization occurring after drainage and reclamation of the former mire ecosystem.

Polish soils are classified as organic if the upper layer contains organic materials thicker than 30 cm, where roots proliferate. Below the root layer, organic soils are further divided according to the thickness of organic materials into shallow (30–80 cm), moderate (80–130 cm), and deep (>130 cm) soils, as well as the nature of the underlying mineral material (Okruszko, 1994). The paludic soils, coded "Pt," characterize peat accumulated in fen, transitional, or bog mire ecosystems. Peat drying and aeration result in the loss of organic substances, and typical soil morphology throughout the organic soil profile. Transformation of the peat materials into post-paludic soils is designed as the MFP (from the Polish name *murszenie*), which is similar to, but more documented than, the muck-forming process (Okruszko, 1960, 1993; van Heuveln et al., 1960; Pons, 1960; Skrynnikova, 1961; Illner, 1977; Schmidt et al., 1981; Zeitz, 1992; Sponagel et al., 1996). The post-paludic soils, coded "Mt," are classified according to the stage of MFP. The moorsh intensity factor is a soil quality indicator for water management (Tomaszewski, 1950; Okruszko, 1960). Proper water management is required not only to fulfill crops' needs, but also to slow down soil subsidence and thus sustain the productive life of cultivated organic soils.

The aim of this chapter is to present the organic soil classification system used in Poland, and the moorsh intensity factor as a quality indicator of reclaimed organic soils.

II. ORGANIC SOIL CLASSIFICATION

In soil classification systems (Agriculture Canada, 1992; Soil Survey Staff, 1996; Schwerdtfeger, 1996; Marcinek, 1997; Sauerbrey and Zeitz, 1999), organic soils are characterized by a histic horizon (Table 1.1), and are genetically related to gleysols (Table 1.2). In Poland, organic soil materials may be peat, mud, mud-peat or gyttja (Table 1.2), depending on the origin of parent materials (Tolpa et al., 1967), as well

Table 1.1 Symbols for Histic Horizons in Different Soil Classification Systems

Histic Horizon	FAO[a]	USA	Germany	England	Russia
Peaty or histic	H	O	H	O	T
Raised bog	H_i	O_i	hH	O_f	T_1
Transitional bog	H_e	O_e	üH	O_m	T_2
Fen	H_a	O_a	nH	O_h	T_3
Ploughed	H_p	O_p	Hp	O_p	T_A

[a]Food and Agriculture Organization.

Source: From Schwaar J. and Schwerdtfeger, G. 1992. *Proc. 9th Int. Peat Congr.*, Uppsala, Sweden, 1:40–45. With permission.

Table 1.2 Comparison between Organic Soil Classification Systems

Polish Classification (1989)	World Reference Base (1994)	FAO-Revised Legend (1990)
IV. HYPROGENIC		
IV A. Paludic soils		
IV A1. Mud		
IVA1a – typic	Eutri – Haplic Organic soils	Hss, Terric Organic soils
IVA1b – peaty	Eutri – Haplic Organic soils	Hss, Terric Organic soils
IVA1c – gyttja	Eutri – Haplic Organic soils	Hss, Terric Organic soils
	Dystri – Haplic Organic soils	Hss, Terric Organic soils
IV A2. Peat		
IVA2a – fen	Eutri – Haplic Organic soils	Eutri-Hss, Eutri-Terric Organic soils, HSt, Fibric Organic soils
IVA2b – transition bog	Dystri – Haplic Organic soils	Dystri-HSf, Dystri-Fibric Organic soils
IVA2c – raised bog	Fibric Organic soils	Dystri-HSf, Dystri-Fibric Organic soils
IV B. Post-paludic soils		
IV B1. Moorsh		
IVB1a – peaty-moorsh	Eutri – Haplic Organic soils	HSs, Terric Organic soils
IVB1b – muddy-moorsh	Eutri – Haplic Organic soils	HSs, Terric Organic soils
IVB1c – gyttja-moorsh	Eutri – Haplic Organic soils	HSs, Terric Organic soils
IVB1d – overmoorsh	Eutri – Haplic Organic soils	HSs, Terric Organic soils
IV B2. Moorsh-like		
IVB2a – mineral-moorsh	Mollic Gleysols	Histi-GLm, Histi-Mollic Gleysols
IVB2b – typic	Areni – Mollic Gleysols	Areni-GLm, Areni-Mollic Gleysols
IVB2c – moorshy (mucky)	Areni – Mollic Gleysols	Areno-GLm, Areni Mollic Gleysols

Source: From Marcinek, J. 1997. *Proc. Comparison between Polish and German Soil Classification Systems*, 13–40. With permission.

as its organic matter content (OMC) (Zawadzki, 1970) and humification degree (Okruszko, 1993). Organic soil materials must show more than 20% OMC. The botanical origin of the peat (Table 1.3), its ash content, and the degree of peat decomposition (Tables 1.4 and 1.5) confer many quality attributes to the soil such as OMC, bulk density, and volume of the solid phase (Tables 1.6 and 1.7) (Ilnicki, 1967;

Table 1.3   Botanical Peat Classification

Type	Subtype	Name
Fen	I. Potamioni	1. Tyrfopel
	II. Limno-Phragmitioni	2. Phragmiteti
	rush peat	3. Scirpo-Typhaeti
		4. Equiseteti
		5. Glycerieti
	III. Magnocaricioni	6. Cariceto-Phragmiteti
	tall sedge peat	7. Cariceti
		8. Cladieti
	IV. Bryalo-Parvocaricioni	9. Bryaleti
	sedge-moss peat	10. Cariceto-Bryaleti
		11. Gramino-Cariceti
	V. Alnioni	12. Saliceti
	alder peat	13. Alneti
		14. Alno-Betuleti
Transition bog	VI. Minero-Sphagnioni	15. Sphagno-Scheuchzerieti
	sedge-sphagnum peat	16. Sphagno-Cariceti
	VII. Betulioni	17. Betuleti
	birch peat	
Raised bog	VIII. Ombro-Sphagnioni	18. Cuspidato-Sphagneti
	sphagnum moss peat	19. Eusphagneti
		20. Eriophoro-Sphagneti
		21. Pino Sphagneti
	IX. Ericioni	22. Ericaceti
	heather peat	23. Trichophoreti
	X. Ledo-Pinioni	24. Pineti
	pine peat	
Peat, nonfibrous	XI. Humotorf	
	sapric peat	

Source: From Tolpa, S., Jasnowski, M., and Palczyski, A. 1967. *Zeszyty Problemowe Postepow Nauk Rolniczych*, 76:9–99. With permission.

Okruszko, 1971; Okruszko and Szymanowski, 1992). Muds are formed under periodic, usually fluvial, flooding. Due to high variations in water level, deposited organic materials are humified, enriched with mineral colloidal materials, and frequently silted. Gyttja is a lacustrine sediment consisting of mineral particles (sand, silt, clay, rests of diatoma, and snails), calcium carbonate, and organic matter (Ilnicki, 1979; Sponagel et al., 1996). They are classified as locustrine chalk, calcareous, calcareous-organic, calcareous mineral, mineral-organic, and organic (Table 1.8).

III. MOORSH FORMATION

The moorsh horizons were first described in the Polish soil classification system (Polish Society of Soil Science, 1974), then in the German scheme. Soils resulting from MFP were called muck soils in the United States, earthy peat soils in Great Britain, and vererdete Torfböden in Germany and The Netherlands (Okruszko, 1998). *The German Soil Cartography Handbook* uses the following terms for the moorsh horizons of organic soils used as meadows (Sponagel et al., 1996): *Torf-Vermulmungshorizont* or peat-dust horizon, *Torf-Vererdungshorizont* or peat-aggregate horizon,

Table 1.4 Peat Classification According to Degree of Decomposition

R Degree	Structure of Peat	Presence of Humus	Amount of Water
R₁ fibric (a)	Spongy or fibrous, occasionally compacted, felty.	None or very little humus, as dispersed dark mass saturating and coloring plant remains.	Large amount of water, easily trickles out, usually almost clear or only slightly brown in color, may contain dark particles of humus.
R₂ hemic (b)	Fibrous- to clody-amorphous, visible fine fibres in humus; peat almost non-elastic after water extraction. Large fragments made of reed or woody remains, crushed between fingers into an amorphous mass.	Humus flowing with water or oozing between fingers and forming less than 1/3 of peat mass.	In paludic sites water oozes in sparse drops or trickles down as a thick, greasy fluid darkened by humus; in drying peat sites, water is slightly colored by humus and may not be used to determine the degree of decomposition.
R₃ sapric (c)	Amorphous structure; dark, homogeneous mass sporadically interspersed with coarser plant remains (wood, reed).	Humus oozing between fingers, comprising at least half of the peat.	Water may not be extracted from peat; mainly humus is extracted.

Source: From Okruszko, H. 1994. *Bibl. Wiadomosci Instytutu Melioracji i Uzytkow Zielonych*, 84:5–27. With permission.

Table 1.5 Comparison between Determination Methods for Degree of Peat Decomposition

Von Post Degree	USSR Humification Degree R	Rubbed Fiber Content			Germany No. in DIN 19682/12	Poland Humification Degree R
		ASTM	USDA-FAO	Canada		
1	R_1	Fibric	Fibric	Fibric	1	R_1
2	R_2	Hemic				
3					2	
4			Hemic	Mesic		R_2
5	R_3	Sapric		3		
6			Sapric			
7					4	R_3
8				Humic		
9					5	
10						

Source: From Malterer, T.J., Verry, E.S., and Erjavec, J. 1992. Proc. 9th Int. Peat Congr., Uppsala, Sweden, 1:310–318. With permission.

Table 1.6 Typical Ranges of Physical Properties of Peat Materials

Type	Most Frequent Occurrence			
	Humification Degree R	Loss on Ignition (kg kg^{-1})	Bulk Density (g cm^{-3})	Volume of Solids (%) (m^3 m^{-3})
Reed peat	R_2 R_3	0.60–0.90	0.15–0.25	0.08–0.11
Tall sedge peat	R_1 R_2 R_3	0.75–0.90	0.10–0.18	0.07–0.10
Sedge-moss peat	R_1 R_2	0.85–0.95	0.09–0.13	0.06–0.08
Alder peat	R_2 R_3	0.70–0.85	0.16–0.22	0.10–0.13
Sedge-*Sphagnum* peat	R_2	0.80–0.95	0.12–0.17	0.07–0.10
Sphagnum peat	R_1	0.97–0.99	0.06–0.10	0.04–0.06

Source: From Ilnicki, P. 1967. Zeszyty Problemowe Postepow Nauk Rolniczych, 76:197–311. With permission.

Table 1.7 Typical Physical Properties of Fen Peat Materials as Related to Degree of Decomposition

Degree	Observations No.	Ash (kg kg^{-1})	Bulk Density (g cm^3)	Porosity	Moisture Content (m^3 m^{-3})		
					pF[a] 2.0	pF 2.7	pF 4.2
R_1	138	0.084	0.114	0.927	0.665	0.396	0.132
R_2	171	0.114	0.141	0.912	0.715	0.496	0.195
R_3	177	0.151	0.175	0.893	0.686	0.527	0.252

[a] pF = $\log_{10}(\psi_M)$, where ψ_M is the matric potential in centimeters.

Source: From Okruszko, H. and Szymanowski, M. 1992. Proc. 9th Int. Peat Congr., Uppsala, Sweden, 3:106–115. With permission.

Table 1.8   Chemical Composition of Gyttja Materials
in Poland

Gyttja Material	$CaCO_3$ Content (g kg^{-1})	Ash Content (g kg^{-1})
Organic gyttja	< 200	< 600
Organic-mineral gyttja	< 200	600–800
Mineral gyttja	< 200	800–1000
Organic-calcareous gyttja	200–600	200–600
Mineral-calcareous gyttja	200–400	600–1000
Calcareous gyttja	400–800	600–1000
Lacustrine chalk	800–1000	800–1000

Source: From Ilnicki, P. 1979. *Proc. Polish Soc. Earth Sci. Conf.,* 73–78. With permission.

and Torf-Bröckelhorizont or peat-lump horizon. Such distinctions were not made in North America, where frequent tillage operations for arable farming admix moorsh horizons. Typically, moorsh thickness is less than 35 cm.

In organic soils, fissures generally start forming at 65% (v/v) moisture content and 70% in the case of alder peat. The *Sphagnum* and sedge-moss fibric peat materials only shrink and do not crack, thus slowing down MFP (Ilnicki, 1967). In practice, MFP develops only in drained fen mires, which make up over 90% of Polish peatlands. As a result of shrinking and swelling, the original peat structure changes into a moorsh structure (Okruszko, 1994). The moorsh grainy or granular structure consists of:

1. Soil aggregates, 5 to 10 mm in diameter, mostly hydrophobic, highly porous, and often made of worm casts
2. Particles not exceeding 4 mm in diameter, highly susceptible to wind erosion, and breaking down easily into smaller fragments

Three moorsh horizons with typical morphological features are recognized in organic soils under permanent grass: M_1 as grainy moorsh in the sod or turf layer and, down the profile, M_2 as humic moorsh, and M_3 as peaty moorsh overlying peat materials (Table 1.9). Compared with original peat materials, moorsh materials show

Table 1.9   General Characteristics of Moorsh Horizons

Horizon	Name	Morphology
M_1	Sod layer: grainy moorsh	Soil mass intermingled with plant roots, structure ranging from amorphous, granular to fine-grained, dust-like. In arable soils, the structure is usually uniform throughout the cultivated layer.
M_2	Subsod layer: humic moorsh	Soil mass characterized by grainy, less frequently granular, relatively loose structure. Soil grains almost exclusively made of compacted humus, their size, between 2–4 mm, gradually increases down the profile to granules 5–10 mm in size.
M_3	Transition layer: peaty moorsh	Soil mass with a peat structure subjected to the moorsh-forming process primarily due to shrinking and swelling leading to the formation of lumps often becoming visible under pressure. The lumps are cemented by humus, frequently leached from overlying layers. Several fissures.

Source: From Okruszko, H. 1994. *Bibl. Wiadomosci Instytutu Melioracji i Uzytkow Zielonych,* 84:5–27. With permission.

Table 1.10 Typical Porosity of Basic Moorsh Types

Type (Layer)	Total Porosity m⁻³	Macroporosity pF < 2.0 m⁻³	Mesoporosity (m³ m⁻³)		Microporosity pF > 4.2 m⁻³
			pF 2.0–4.2 m⁻³	pF 2.0–2.7 m⁻³	
Grainy (M_1)	0.825	0.249	0.291	0.122	0.285
Humic (M_2)	0.830	0.172	0.382	0.186	0.276
Peaty (M_3)	0.885	0.161	0.507	0.257	0.217

Source: From Okruszko, H. 1976. *Zeszyty Problemowe Postepow Nauk Rolniczych,* 177:159–204. With permission.

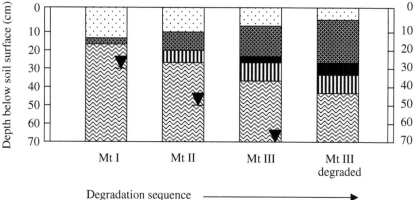

▼ Average depth of groundwater table cross vegetation period

Figure 1.1 Organic soil profiles at four developmental stages of the moorsh-forming process. (Adapted from Okruszko, H. 1967. *Zeszyty Problemowe Postepow Nauk Rolni-czych,* 72:13–27. With permission).

15–20% higher ash content, 0.2–0.3 g cm⁻³ higher bulk density, and smaller porosity (Table 1.10). The retention of plant available water is considerably reduced in grainy moorsh compared with original peat materials (Table 1.10), whereas hydraulic conductivity is highest and capillarity rise is lowest in the grainy moorsh compared with original peat materials. The thicker the grainy moorsh horizon (M_1), the lower the soil quality is. The M_3 horizon may include a 5- to 10-cm thick degraded layer of coke-like peat materials made of hard, small, and irreversibly dried particles (Figure 1.1), indicating markedly reduced soil quality.

IV. SOIL MORPHOLOGY AS AN INDICATOR OF ORGANIC SOIL QUALITY

The soil profile is divided into peat and moorsh horizons (Table 1.11). Three layers are examined to assess soil processes across a 130-cm deep soil profile (Okruszko, 1994):

1. 0–30 cm: attributes of the root layer (K), primarily water retention capacity
2. 30–80 cm: attributes of the underlying layer (T_1) such as water storage capacity and capillary rise for transferring water from water table to the root zone
3. 80–130 cm: the lower layer (T_2) affecting lateral water movement connected with drainage ditches or groundwater flow

The soil process is identified as P for paludic soils in mire ecosystems, or M for moorsh or post-paludic soils in peatlands. Mire paludification degree is designed from the degree of peat decomposition as fibric (III), hemic (II), and sapric (I) peat materials. The symbol "t" originates from the Polish and German word *Torf* (peat). In general, highly paludified soils (PtIII) (i.e., fibric peat materials) change into low-degree moorsh (MtI), while slightly to moderately paludified soils (PtI to PtII) change into medium-degree moorsh (MtII). The thickness of genetic moorsh horizons qualifies MFP by degradation stage (Figure 1.1). The intensity of MFP (I to III in the ascending order of peat degradation rate) depends on thickness of the M_1-M_2-M_3 horizon sequence in K (Tables 1.9 and 1.12). The three degrees of decomposition in T_1 (30–80 cm) are: a for fibric peat, b for hemic peat, and c for sapric peat. Soil notation then becomes, for example, MtIIb. Layer T_2 (80–130 cm) is described as layer T_1, for example, MtIIbc. These rules are extended to deeper layers (>130 cm).

Polish fen soils are typically thicker than 80 cm. In moderately deep soils (80–130 cm), mineral layers underlying peat should be characterized for grain-size distribution (Table 1.13). For example, a moderately deep organic soil may be coded as MtIIb2.

V. PROGNOSTIC SOIL MOISTURE COMPLEXES (PSMC) AS RELATED TO ORGANIC SOIL QUALITY

After organic soil reclamation, capillary rise supplying water to the roots is related to peat structure. Many soil types from various combinations of diagnostic horizons (Okruszko, 1977, 1994) were grouped in prognostic soil moisture complexes (PSMC). For managing Polish fen soils under grassland farming, five PSMCs were named after typical water regimes as wet (A), periodically wet (AB), moist (B), periodically dry (BC), and dry (C). Soil types in each PSMC are assumed similar in water regime and agricultural use as shown in Table 1.14 (Szuniewicz, 1994).

Table 1.11 Scheme for Describing a Moorsh Soil Profile

Diagnostic Layer	Classification
K (0–30 cm)	Soil group as paludic Pt or moorsh Mt, then as PtT-II-III or MtI-II-III (Table 1.12)
T_1 (30–80 cm)	Soils are classified: a, b, c (Table 1.4)
T_2 (80–130 cm)	Soils are classified as (Tables 1.4 and 1.13): aa-ab-ac or a1-a2-a3-a4 ba-bb-bc or b1-b2-b3-b4 ca-cb-cc or c1-c2-c3-c4

Source: From Okruszko, H. 1998. *J. Water Land Dev.*, 2:63–73. With permission.

Table 1.12 Stage of Soil Degradation Due to the Moorsh-forming Process

Stage	Genetic Layers	Thickness of the Moorsh	Features of the Moorsh
MtI (initial)	M_1 M_2 beginning	<20 cm	Usually poor granulation, M_1 strongly bound by roots, peat fibres in the moorsh
MtII (intermediate)	M_1 and M_2 M_3 beginning	20–30 cm	Thick moorsh, abrupt transition from moorsh to underlying parent material, no well-defined M_3
MtIII (final)	M_1, M_2 and M_3	>30 cm	Well-defined M_3 with coarse, prismatic or fissured structure, dehydration under the root zone

Source: From Okruszko, H. 1994. *Bibl. Wiadomosci Instytutu Melioracji i Uzytkow Zielonych*, 84:5–27. With permission.

Table 1.13 Textural Classification of Mineral Soils in Poland

Textural Type	Symbol	Granulometric Classification
Very light	1	loose sand, slightly loamy sand
Light	2	light loamy sand, heavy loamy sand
Medium	3	light loam, medium loam, silt
Heavy	4	loamy silt, heavy loam, clay

Source: From Okruszko, H. 1998. *J. Water Land Dev.*, 2:63–73. With permission.

The wet complex (A) occurs in slightly decomposed sedge-moss peat and represents areas receiving a liberal groundwater supply. Due to the fibrous peat structure conferring high capillary rise, the soil is moist even when the groundwater level is 100 cm below surface. After drainage, they change slowly during the first moorsh stage (MtI). The rate of decomposition is slow due to low available N. The periodically wet complex (AB) resembles the A complex: the fibric sedge-moss peat overlies hemic, usually tall sedge peat, resulting in lower capillary rise and leading to smaller moisture acquisition by the roots in periods of intensive evapotranspiration.

The moist complex (B) occurs in hemic, usually tall sedge peat deposits, occasionally showing fibric peat in deeper sections of the profile. Due to high volume of mesopores, sufficient moisture is available for grass growth if the groundwater table is lowered no deeper than 80 cm. After several years of drainage and droughts, capillary rise and water retention capacity decrease.

The periodically dry complex (BC) develops on soils formed from hemic or fibric peat materials overlying sapric peat, mud, or gyttja, as well as medium- to heavy-textured mineral soil materials. By lowering the groundwater table, the surface layers may become too dry to sustain grass growth. Organic materials undergo intensive MFP and release high amounts of mineral N. The drying complex (C) is associated with shallow soils developed from sapric peat overlying light- or heavy-textured soil materials. The soil becomes too dry in the summer without supplementary irrigation, because water level is too low for effective capillary rise.

Table 1.14 Classification of Prognostic Soil-Moisture Complexes in Peatlands under Grassland in Poland

Prognostic Soil-Moisture Complex	Thickness of Organic Soils		
	Thick (>130 cm)	Moderately Thick (80–130 cm)	Shallow (<80 cm)
A (wet)	Mtlaa		
AB (periodically wet)	Mtlab Mtlba	Mtllaa	
B (moist)	Mtlbb Mtllab Mtllba Mtllbb	Mtlaa Mtlab Mtlba Mtlbb	Mtla3
BC (periodically dry)	Mtlac Mtlbc Mtlllba	Mtllab Mtllac Mtllba Mtllbb Mtllbc	Mtla4 Mtlb3 Mtla3 Mtlb3
C (dry)	Mtllca Mtllcb Mtllcc Mtlllca Mtllbb Mtllbc Mtlllcb Mtlllcc	Mtllba Mtllbb Mtllbc Mtllca Mtllcb Mtllcc Mtla2 Mtla1 Mtlb4 Mtlb2 Mtlb1 Mtlla4 Mtlla2 Mtlla1 Mtllb4 Mtllb2 Mtllb1	Mtllc3 Mtllc4 Mtllc2 Mtllc1 Mtlllb3 Mtlllb4 Mtlllb2 Mtlllb1 Mtlllc3 Mtlllc4 Mtlllc2 Mtlllc1

Source: From Okruszko, H. 1998. *J. Water Land Dev.*, 2:63–73. With permission.

The following reclamation and water management measures regulate air–water conditions:

1. For A and AB, irrigation is not needed.
2. The B complex requires careful regulation of water outflow in order to keep water easily available to the plant under the root zone, and to prevent degradation of soil physical properties such as capillary rise and water retention capacity.
3. For the BC and C complexes, drainage and irrigation must be combined to regulate water outflow for sites supplied by groundwater inflow. If groundwater inflow is low, irrigation water is directed to the drainage system from reservoirs or streams. If shallow organic soils are underlain by porous mineral substrate, the water level is raised in the river to facilitate lateral water inflow through the permeable substrate.

VI. CONCLUSION

The intensity of MFP taking place in drained organic soils is a valuable indicator of soil quality. Strong moorsh development indicates degraded physical properties and increased N mineralization potential. To describe MFP, the Polish soil classification system defined three diagnostic layers (0–30, 30–80, 80–130 cm), three stages of advancement of the moorsh-forming process in the root layer (I, II, III), three characteristic genetic horizons (M_1, M_2, M_3), and three peat decomposition degrees (a, b, c) in the two underlying peat layers (T_1 and T_2). Those physical and morphological indicators are used to classify soils. Soil types are further grouped into five prognostic soil moisture complexes for managing the water resource in drained organic soils (A, AB, B, BC, and C).

REFERENCES

Agriculture Canada. 1992. Canadian system of soil classification. (in French). Publ. 1646, 2nd Ed., Research Branch, Agriculture and Agri-Food Canada, Ottawa.

Illner, K. 1977. Soil genesis in fen peat. (in German). *Archiv für Acker Pflanzenbau und Bodenkunde*, 21(12):867–782.

Ilnicki, P. 1967. Peat shrinkage by drying as related to structure and physical properties. (in Polish with German summary). *Zeszyty Problemowe Postepow Nauk Rolniczych*, 76:197–311.

Ilnicki, P. 1979. Prospects for utilizing calcareous gyttja as commercial fertilizer. (in Polish). *Proc. Polish Soc. Earth Sci. Conf.*, 73–78, Kreda jeziorna i gytia, Lubniewice-Gorzów.

Malterer, T.J., Verry, E.S. and Erjavec, J. 1992. Peat classification in the relation to several methods used to determine fiber content and degree of decomposition. *Proc. 9th Int. Peat Congr.*, 1:310–318, Uppsala, Sweden.

Marcinek, J. 1997. Principles of the Polish soil classification system. *Proc. Comparison between Polish and German Soil Classification Systems*: 13–40, German Soil Sci. Soc. and Polish Soil Sci. Soc., Szczecin, Poland.

Okruszko, H. 1960. Moorsh soils of valley peatlands, and their chemical and physical properties. (in Polish with English summary). *Roczniki Nauk Rolniczych*, F 74.1: 5–89.

Okruszko, H. 1967. Soil properties in drained peatland. (in Polish with German summary). *Zeszyty Problemowe Postepow Nauk Rolniczych*, 72:13–27.

Okruszko, H. 1971. Determination of specific gravity of hydrogenic soils on the basis of their mineral particles content. (in Polish with English summary). *Wiadomosci Instytutu Melioracji i Uzytkow Zielonych*, 10.1:47–54.

Okruszko, H. 1976. Impact of draining organic soils under Polish conditions. (in Polish with English summary). Zeszyty Problemowe Postepow Nauk Rolniczych, 177:159–204.

Okruszko, H. 1977. Prognostic soil-moisture complexes in drained peatlands as criterion for differentiating site conditions. (in Polish with English summary). *Zeszyty Problemowe Postepow Nauk Rolniczych*, 186:49–66.

Okruszko, H. and Szymanowski, M. 1992. Correlation between bulk density and water-holding capacity of fen peats. *Proc. 9th Int. Peat Congr.*, 3:106–115, Uppsala, Sweden.

Okruszko, H. 1993. Transformation of a fen organic soil after drainage. *Zeszyty Problemowe Postepow Nauk Rolniczych*, 406:3–73.

Okruszko, H. 1994. System for hydrogenic soil classification in Poland. Bibl. *Wiadomosci Instytutu Melioracji i Uzytkow Zielonych*, 84: 5–27.

Okruszko, H. 1998. System of fen-peat soils description used in Poland. *J. Water Land Dev.*, 2:63–73.

Polish Society of Soil Science. 1974. Soil genesis, classification and cartography. Com. V systematics of Polish soils. *Roczniki Gleboznawcze*, 25:1–148.

Pons, L.J. 1960. Soil genesis and classification of reclaimed peat soils in connection with initial soil formation. *Trans. 7th Int. Congr. Soil Sci.*, 4:205–211, Madison, WI.

Sauerbrey, R. and Zeitz, J. 1999. Peatlands. (in German). Section 3.3.3.7 in H.P. Blume (Ed.). *Handbuch der Bodenkunde*. Losebllätter Ausgabe, Ecomed, Landsberg, Germany.

Schmidt, W., Scholz, A., Mundel, G., and van der Waydbrink, W. 1981. *Characteristics and Assessment of Soil Development in Fen Peat as Related to Degradation*. (in German). Forschungsbericht Institut für Futterproduktion Paulinenaue, Akademie der Landwirtschaftlichen Wissenschaften DDR, Berlin.

Schwaar, J. and Schwerdtfeger, G. 1992. Peat horizons and their use to typify mire profiles. *Proc. 9th Int. Peat Congr.*, 1:40–45, Uppsala, Sweden.

Schwerdtfeger, G. 1996. Soil classification for mires. *Proc. 10th Int. Peat Congr.*, 2:60–66, Bremen, Germany.

Skrynnikova, I.N. 1961. *Soil Processes in Cultivated Peat Soils*. (in Russian). Izdatielstvo Akademii Nauk SSSR, Moscow, Russia.

Soil Survey Staff. 1996. *Keys to Soil Taxonomy*. SMSS. Technical Monograph No. 19, 7th Ed., Pocahontas Press Inc., Blacksburg, Virginia.

Sponagel, H., et al. 1996. *Methods of Soil Cartography*. (in German). E. Schweizerbart'sche Verlagsbuchhandlung (Nägele u. Obermiller), Stuttgart, Germany.

Szuniewicz, J. 1994. Characteristics of prognostic soil-moisture complexes in terms of parameters of melioration system. *Bibl. Wiadomosci Instytutu Melioracji i Uzytkow Zielonych*, 84:35–57.

Tolpa, S., Jasnowski, M., and Palczyski, A. 1967. System of genetical classification of peat in middle Europa. (in German). *Zeszyty Problemowe Postepow Nauk Rolniczych*, 76:9–99.

Tomaszewski, J. 1950. Conditions of genesis, development and metamorphosis of swampy soils. (in Polish with English summary). *Roczniki Nauk Rolniczych*, 54:607–629.

van Heuveln, B., Jongerius, A., and Pons, L.J. 1960. Soil formations in organic soils. *Trans. 7th Int. Congr. Soil Sci.,* 4:195–204, Madison, WI.

Zawadzki, S. 1970. Relationship between the content of organic matter and physical properties of hydrogenic soils. *Pol. J. Soil Sci.,* 3:3–9.

Zeitz, J. 1992. Physical properties of soil horizons in fen organic soils used in agriculture. (in German). *Zeitschrift für Kulturtechnik und Landentwicklung,* 33:301–307.

CHAPTER 2

Irreversible Loss of Organic Soil Functions after Reclamation

Piotr Ilnicki and Jutta Zeitz

CONTENTS

ABSTRACT

After drainage, organic soils change their basic functions from natural carbon sinks and water reservoir to sources of greenhouse gases and water-deficient bodies. The natural process of carbon sequestration is paludification; with drainage and aeration, the organic soil undergoes the irreversible moorsh-forming process (MFP). The intensity of MFP is shown by morphological and structural transformations, enrichment in humic substances, changes in mineral composition, as well as shifts in microbial populations, mesofauna and earthworm species. The climatic impact factor (CO_2 + CH_4 + NO_x) of organic soil cultivation would be between 2.9 and 10.3 Mg CO_2 ha^{-1} yr^{-1}. Maximum CO_2 production is associated with arable farming and 90-cm deep water table level. The easily mineralizable N pool makes up 0.4 to 2.8% of total N in the 0–20 cm layer, supplying 77 to 493 kg N ha^{-1} yr^{-1} as mineral N depending on moorsh stage. Optimum volumetric air content for N mineralization is 20–30%. There is 20% more N mineralized under arable farming compared with grassland. The NO_3-N to NH_4-N ratio increases with MFP, thus enhancing N leaching and denitrification in anaerobic microsites. Addition of N-bearing fertilizers increases N pollution hazards. Organic soil quality as monitored by MFP attributes is best maintained under grassland farming with high groundwater level.

I. INTRODUCTION

Drainage must increase volumetric air content to at least 6–8% in the upper layer of organic soils used as grassland (Okruszko, 1993). Air contents up to 20–30% provide optimum conditions for intensive MFP, the transformation of peat materials into moorsh. The MFP is initiated by soil consolidation and subsidence after drainage, then accelerated by repeated shrinkage and swelling upon successive drying and wetting, and by microbial decomposition of organic substances. The peat mineralization rate depends on degree of decomposition and ash content, temperature, water and air contents, and nutrient ratios. It is faster in fen than in oligotrophic or bog peats, and in soils used for arable farming compared with grassland.

The MFP, the reverse process of paludification, was defined by Okruszko (1985) as decession (from the Latin word *decessio* meaning loss or dissipation). The MFP leads to the gradual disappearance of organic soils from the landscape. The MFP contributes to CO_2 emissions depending on intensity, and is associated with irreversible transformations of peat properties as driven by drier soil conditions. Peatland functions for conserving water and as carbon sink are thus drastically reversed following drainage. Monitoring peat properties during MFP helps planning soil utilization and conservation.

The aim of this chapter is to present organic soil indicators of the decrease in organic soil quality following drainage and reclamation.

II. MORPHOLOGICAL CLASSIFICATION OF GENETIC SOIL HORIZONS

With the decrease in water content after drainage, peat structure changes gradually to a more or less crumby, granular or grainy structure (Okruszko, 1993). Throughout the surface layer, the peat mass is fragmented into a fine, sometimes dusty material, due to MFP. The size of moorsh particles increases with soil depth. A morphological classification system for MFP was first proposed by Okruszko (1960, 1993, 1994) in Poland, followed by Schmidt and Illner (1976) in Germany. A comparative nomenclature of moorsh horizons in grassland soils is presented in Table 2.1.

The characteristic moorsh horizons are genetically related to one another in the soil profile as a result of gradual transformation of soil physical, chemical, and

Table 2.1   Symbols used in Poland and Germany for Designating the Moorsh Horizons of Drained Organic Soils

Poland[a]		Germany[b]	
Symbol	Layer Morphology	Symbol	Layer Morphology
M_1 grainy moorsh	At sod level, soil mass bound by plant roots, structure ranging from granular to fine-grain, dust-like. In arable soils, the structure is usually uniform throughout the cultivated layer.	Hm	*Torf-Vermulmungshorizont* (peat-dust horizon) at the surface of intensively drained and tilled organic soils, high degree of decomposition when dry; very fine granular and dusty, high water repellency.
M_2 humic moorsh	Under sod, soil mass characterized by grainy, less frequently granular, relatively loose structure. Soil grains made of compacted humus. Their size is 2–4 mm, gradually increasing down the profile to 5–10 mm.	Hv	*Torf-Vererdungshorizont* (peat-earth horizon), low to moderate humification, crumby or fine subangular structure.
M_3 peaty moorsh	Transitional horizon, soil mass with a peat structure subjected shrinkage and swelling, producing lumps or aggregates often visible under pressure. The lumps are cemented by humus, frequently leached from overlying layers. Several fissures.	Ha	*Torf-Bröckelhorizont* (peat-crumb horizon), coarse to fine-angular blocky structure, vertical and horizontal shrinkage cracks.
		Ht	*Torf-Schrumpfungshorizont* (peat shrinkage horizon), vertical cracks and coarse prismatic structure caused by shrinkage.
T_1 peat layer	Underlying peat horizon above groundwater level.	Hw	Horizon affected by fluctuating groundwater or perched-water table, partially oxidized.
T_2 peat layer	Underlying peat horizon below groundwater level.	Hr	*Torf-Horizont* (peat horizon) below groundwater table, reduced state

[a] *Source:* From Okruszko, H. 1993. *Pol. Akad. Nauk,* 406:3–75; Okruszko, H. 1994. *Bibl. Wiadomosci Instytutu Melioracji i Uzytkow Zielonych,* 84:5–27. With permission.
[b] *Source*: Sponagel, H., et al. 1996. *Methods of Soil Cartography* (in German). E. Schweizerbart'sche Verlagsbuchhandlung (Nägele u. Obermiller), Stuttgart, Germany, 392 pp.; Schäfer, W. 1996. *Proc. 10th Int. Peat Congr.,* 4:77–84. With permission.

biological properties after drainage. Physical properties of the more humified horizon Hm or M_1 differ markedly from those of the less humified Hv or M_2. From a soil conservation viewpoint, the Hm to Ha or M_1 to M_3 horizon sequences are indicative of the degree of soil degradation through MFP.

III. CHANGES IN PHYSICAL PROPERTIES

A. Peat Shrinkage

Soil volume losses between 53 and 70% compared with the initial peat volume vary with peat botanical composition and degree of decomposition (Table 2.2). Peat shrinkage is larger the higher the degree of decomposition and the smaller the ash

Table 2.2   Relationship between Peat Shrinkage and Botanical Composition

Peat Type	No. of Samples	Shrinkage $(m^3 \, m^{-3})$	DD^a (%)	Ash Content $(kg \, kg^{-1})$	Bulk Density $(g \, cm^{-3})$	Solid Phase $(m^3 \, m^{-3})$
Reed	25	0.65	45	0.306	0.204	0.109
Sedge-reed	22	0.68	40	0.227	0.196	0.109
Sedge	29	0.70	40	0.176	0.156	0.090
Moss	15	0.59	29	0.144	0.121	0.071
Sedge-moss	24	0.60	28	0.135	0.132	0.077
Alder	27	0.70	53	0.260	0.212	0.117
Sphagnum	6	0.53	13	0.041	0.069	0.040
Sphagnum	13	0.66	40	0.026	0.116	0.072

^{a} Degree of peat decomposition in % (Russian method).

Source: From Ilnicki, 1967. *Zeszyty Problemowe Postepow Nauk Rolniczych,* 76: 197–311. With permission.

content. An increased moss fraction decreases peat shrinkage. A linear relationship exists between volumes of total (Y in %) and of irreversible shrinkage (X in %) as follows (Ilnicki, 1967):

$$Y = 0.83X - 4.70, \quad R^2 = 0.67 \tag{2.1}$$

Irreversible shrinkage after drainage is a characteristic of MFP. The higher the peat decomposition and the more advanced the peat drying, the greater the structural changes through MFP. Fissures starting to develop at 65 to 75% volumetric moisture content become obvious at 50% moisture content (Ilnicki, 1967).

B. Peat Density

Bulk and particle densities are parameters of soil porosity. They change during MFP due to compaction and increased ash content. Average particle density of peat organic matter is 1.45 g cm^{-3}, varying from 1.3 to 1.6 g cm^{-3} (Okruszko, 1993). Particle density (PD) depends on ash content (Table 2.3). Linear relationships between PD (g cm^{-3}) and ash content (% w/w) were described as follows for peat materials:

$$PD = 0.011ash + 1.45 \tag{2.2}$$

(Okruszko, 1971)

$$PD = 0.0086ash + 1.44 \tag{2.3}$$

(TGL 31222/03, 1985)

and for mud materials:

$$PD = 0.0124ash + 1.35 \tag{2.4}$$

(TGL 31222/03, 1985)

Table 2.3   Properties of 1470 Organic Soil Materials from Poland with Varying Ash Contents

Peat Material	Ash Content (kg kg⁻¹)	Particle Density (g cm⁻³)	Bulk Density (g cm⁻³)	Solid Phase (m³ m⁻³)	Pore Volume (m³ m⁻³)
Unsilted	0.05–0.25	1.51–1.73	0.11–0.19	0.07–0.11	0.89–0.93
Silted	0.25–0.50	1.73–2.00	0.19–0.29	0.11–0.16	0.84–0.89
Strongly silted	0.50–0.80	2.00–2.33	0.29–0.41	0.16–0.22	0.78–0.84

Source: From Okruszko, 1976. *Bibl. Wiadomosci Instytutu Melioracji i Uzytkow Zielonych*, 52:7–54. With permission.

Table 2.4   Volume of the Solid Phase in Moorsh Materials

Moorsh Stage	Horizon	Depth (cm)	Samples (No.)	Ash (kg kg⁻)	Bulk Density (g cm⁻³)	Solid Phase (m³ m⁻³)
MtI	M_1	0–10	45	0.160	0.203	0.127
	M_2	10–20	28	0.145	0.179	0.112
	M_3	20–30	10	0.122	0.160	0.102
	T_1	40–60	53	0.104	0.143	0.096
	T_2	80–100	10	0.101	0.126	0.084
MtIII	M_1	0–10	41	0.176	0.321	0.192
	M_2	10–20	41	0.156	0.298	0.180
	M_3	20–30	62	0.124	0.230	0.142
	T_1	40–60	69	0.108	0.155	0.097
	T_2	80–100	10	0.102	0.134	0.084

Source: From Okruszko, 1976. *Bibl. Wiadomosci Instytutu Melioracji i Uzytkow Zielonych*, 52:7–54. With permission.

The volume of peat solid phase, calculated as the ratio of bulk density to particle density (Okruszko, 1993), increases with ash content and degree of decomposition (Tables 2.2 and 2.3). Drainage and soil drying increase bulk density and volume of the solid phase in the upper layer (Table 2.4). The volume of solids in fen organic soils from Germany increased by 100 to 300% in the top layer, and by 50 to 100% in the subsoil, after 27 years of MFP (Figure 2.1). The volume of solids is thus a useful indicator of MFP.

C. Peat Porosity

Peat porosity ranges between 78 and 93%. The higher the degree of decomposition, the larger the volume of micropores, and the smaller the volume of macropores and mesopores in peat materials will be. The MFP alters porosity, pore size distribution, and soil water regime (Tables 2.5 and 2.6). The MFP in fen peats decreased total porosity by 3% (peaty moorsh) to 9% (grainy moorsh). The volume of macropores and micropores increased at the expense of mesopores (Table 2.6). Transition from peat to moorsh decreased water availability to plants. Porosity of peaty moorsh materials was similar to hemic peats, and grainy moorsh resembled sapric peats (Okruszko, 1993). Volume of micropores (<0.2 μm) and larger macropores (>300 μm) increased at the expense of smaller macropores (300–30 μm)

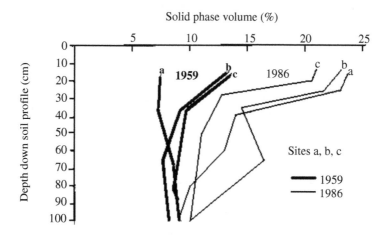

Figure 2.1   Percentage volume change in the solid phase in three thick organic soils between 1959 (Titze, Water and air composition of the upper earthy layer of the Klenzer fen and its influence on yield, Universite Rostock, 1966 in bold characters) and 1986 (Zeitz, *Zeitschrift für Kulturtechnik und Landentwicklung*, 32:227–234, 1992).

Table 2.5   Average Porosity of Peat and Moorsh Materials in Poland

Peat or Moorsh Material	Porosity	Macroporosity pF < 2.0 (m³ m⁻³)	Mesoporosity pF 2.0–2.7 (m³ m⁻³)	pF 2.7–4.2 (m³ m⁻³)	Microporosity pF > 4.2 (m³ m⁻³)
Moss-sedge peat R_1	0.920	0.257	0.307	0.533	0.132
Alder swamp peat R_3	0.885	0.248	0.145	0.352	0.207
Peaty moorsh	0.885	0.161	0.257	0.507	0.217
Humic moorsh	0.830	0.172	0.186	0.382	0.276
Grain moorsh	0.825	0.249	0.122	0.291	0.285

Source: From Okruszko, 1993. *Pol. Akad. Nauk*, 406: 3–75. With permission.

as MFP advanced (Burghardt and Ilnicki, 1978). Hysteresis was found to be larger in peat than in moorsh materials, peaty soil or humic sand (Ilnicki, 1982). Suctions varying from −1 to −3 kPa caused differences of 0.076 cm³ cm⁻³ in water content during peat drying and rewetting cycles. Hysteresis of water retention curves was smaller the higher the degree of peat decomposition, MFP intensity, ash content, bulk density, and pH. Hysteresis increased with the volume of macropores (>50 μm).

D. Hydraulic Conductivity

In the saturated zone of the peat profile, hydraulic conductivity (k_f) generally decreases with time and drainage intensity due to peat compaction. Preferential flow increases with shrinkage fissures in the moorsh compared with peat layers (Table 2.7). The more advanced the MFP, however, the lower was the hydraulic conductivity (Zeitz, 1991; Sauerbrey and Zeitz, 1999).

Table 2.6   Properties (Mean ± Standard Deviation) of Moorsh Horizons in Oligotrophic Organic Soils

Horizon	v. Post H	Ash (kg kg⁻¹)	Bulk Density (g cm⁻³)	TP[a] (m³ m⁻³)	VSP[a] (m³ m⁻³)	FC[a] (m³ m⁻³)	AP[a] (m³ m⁻³)	PAW[a] (m³ m⁻³)
nHm	7 ± na[b]	0.29 ± 0.04	0.28 ± 0.2	0.82	0.18 ± 0.03	0.68 ± 0.07	0.14 ± 0.07	0.36 ± 0.16
nHv	7 ± 1.5	0.19 ± 0.20	0.03 ± 0.06	0.86	0.14 ± 0.11	0.70 ± 0.13	0.16 ± 0.07	0.46 ± 0.14
nHa nHt			Not occurring in fibric peat materials					
nHw H3-H4	3 ± 0.4	0.03 ± 0.013	0.10 ± na	0.94	0.06 ± 0.012	0.73 ± 0.06	0.12 ± 0.06	0.57 ± 0.06
H5-H6	5 ± na[b]	0.02 ± 0.005	0.10 ± 0.02	0.94	0.06 ± 0.014	0.76 ± 0.07	0.18 ± 0.12	0.58 ± 0.08
H7-H8	8 ± 0.5	0.02 ± 0.013	0.12 ± 0.02	0.92	0.08 ± 0.009	0.80 ± 0.04	0.12 ± 0.05	0.60 ± 0.03
nHr H3-H4	3 ± 0.4	0.02 ± 0.001	0.07 ± 0.03	0.96	0.04 ± 0.004	0.67 ± 0.03	0.29 ± 0.04	0.59 ± 0.04
H5-H6	5 ± 0.4	0.01 ± 0.006	0.14 ± 0.02	0.91	0.09 ± 0.009	0.82 ± 0.03	0.08 ± 0.03	0.64 ± 0.03
H7-H8	8 ± 0.6	0.04 ± 0.047	0.13 ± 0.04	0.92	0.08 ± 0.021	0.83 ± 0.02	0.08 ± 0.02	0.62 ± 0.08

[a] TP = total porosity; VSP = volume of the solid phase; FC = water at field capacity (pF > 1.8); AP = air porosity at field capacity (AP = TP − VSP − FC); PAW = plant available water between pF 1.8 (field capacity) and 4.2 (wilting point).
[b] na = not available.

Source: From Schäfer, 1996. Proc. 10th Int. Peat Congr., Bremen, Germany, 4:77–84. With permission.

Table 2.7   Permeability Change in Soil Layers across a Peat-Moorsh Soil Profile

Layer	Vertical (cm d⁻¹)	Lateral (cm d⁻¹)	Mean (cm d⁻¹)
	Saturated Hydraulic Conductivity (Average)		
Moorsh M_1	160	73	104
Moorsh M_2	136	59	82
Moorsh M_3	61	31	42
Peat T_1	7	5	6
Peat T_2	8	5	7

Source: From Okruszko, 1960. *Roczniki Nauk Rolniczych*, F74:5–89. With permission.

Table 2.8   Unsaturated Hydraulic Conductivity as Related to Moorsh and Peat Layers in the Upper Rhinluch Peatland, Germany

Depth (cm)	Horizon (Symbol)	pF 1.5 (mm d⁻¹)	pF 1.8 (mm d⁻¹)	pF 2.0 (mm d⁻¹)	pF 2.2 (mm d⁻¹)	pF 2.5 (mm d⁻¹)
		Unsaturated Hydraulic Conductivity				
0–10	nHm	2.010	1.394	0.294	0.059	0.008
20–30	nHv	2.934	1.573	0.323	0.077	0.015
30–40	nHa	4.715	2.078	0.468	0.144	0.037
50–60	nHt	6.324	3.114	0.828	0.227	0.056
70–80	nHt	4.874	2.810	0.540	0.142	0.004

Source: From Sauerbrey and Zeitz, 1999. Peatlands. Section 3.3.3.7, in *Handbuch der Bodenkunde*. (in German). Blume, H.P., Ed., *Loseblätter Ausgabe*, Ecomed Publ., Landsberg, Germany, 20 pp. With permission.

Capillary rise, which depends on mesopores in the peat layer underlying the moorsh, is lower when the degree of decomposition is higher, and varies in height from 70 to 160 cm. Rate of capillary rise for fibric peats during intensive evapotranspiration reaches 10 mm per day (Baden and Eggelsmann, 1963; Szuniewicz and Szymanowski, 1977). Comparatively, capillary rise would cover 56% of the evapotranspiration demand in hemic peats and 17 to 23% in sapric peats, depending on the advancement of MFP. Unsaturated hydraulic conductivity decreases sharply in moorsh compared with peat materials (Table 2.8). The height of capillary rise would be less than 10 cm in deeper moorsh layers. Therefore, groundwater must be maintained at a higher level for grassland grown in those deep moorsh soils (60 cm in hemic peats and 30 to 50 cm in sapric peats, depending on MFP).

The unit water content (UWC) is a rough indicator of structural changes in peat. The UWC is the relative volumetric water content of a disturbed peat sample before and after consolidation under a pressure of 100 kPa. The more advanced the MFP, the lower are water retention capacity and UWC. For organic soils in an advanced stage of MFP, the structure is similar to a single-grain mineral soil. The UWC exceeds 2.2 for low MFP, and is less than 1.5 for high MFP. The UWC of the Hm horizon is 20% lower than that of Hv (Zeitz and Tölle, 1996), thus indicating a higher degree of MFP for Hm.

IV. CHANGES IN CHEMICAL PROPERTIES

Peat drainage affects the composition of organic materials, the mineralization of nitrogen and carbon (emission of greenhouse gases), the composition of inorganic materials, and drainage water quality. The organic material undergoing humification comprises bitumens, hemicellulose, lignin, and humic substances (Okruszko, 1993). Moorsh materials are particularly enriched in humic substances and impoverished in lignin compared with original peat materials (Table 2.9).

Luthardt (1987) and Behrendt (1995) found that:

1. Cellulose decomposition rate was smaller in MtI-MtII than in MtIII
2. Tillage promoted cellulose decomposition
3. Cellulose decomposition was correlated to volumetric soil moisture content with maximum rate at 70%
4. Cellulose decomposition rate was higher in intensively cultivated fen soils compared to unplowed areas

In moorsh materials, the ratio of humic to fulvic acids is smaller than in the original peat materials (Okruszko, 1993). The moorsh materials are enriched in inorganic materials such as Si, Fe, P, and Al compared with the original peat materials (Table 2.10). Microelements, sorbed by colloidal organic matter, accumulate in the moorsh layer (Okruszko, 1993).

More mineralizable N is present in moorsh than in peat materials. Enhanced N mineralization by 30% in moorsh compared with peat increases N availability to plants, nitrate leaching potential, and N loss through denitrification. Nitrate leaching is at risk for drinking water, while denitrification may evolve NO_x gases, which contribute to global climate changes. The amount of N bound to fulvic acids, hemicellulose and cellulose is increased by 30% in moorsh compared with peat materials (Okruszko, 1993). The most easily mineralized N pool makes up 0.4 to 2.8% of total moorsh N, supplying 77 to 493 kg N ha^{-1} yr^{-1} as mineral N in the 0–20-cm layer, depending on moorsh stage (Table 2.11). Highest N mineralization rate in the moorsh layers occurred 5–10 cm below soil surface in the spring, and 15–20 cm below surface in the summer (Frackowiak, 1969). Optimum volumetric air content in the soil for N mineralization is 20–30%. There was 20% more N mineralized under arable farming compared with grassland (Gotkiewicz et al., 1975). The NO_3-N to NH_4-N ratio increased with MFP and soil aeration (Gotkiewicz and Szuniewicz, 1987). The application of mineral fertilizer reduced N mineralization. Very high application rates of mineral N (e.g., 480 kg N ha^{-1} as calcium ammonium nitrate) and organic N (371 kg N ha^{-1} as cattle slurry) may cause a short-term increase in NO_x emissions (Augustin, 2001).

The rate of organic matter decomposition is usually assessed from CO_2 evolution. Decay rate of organic C is smaller when the degree of decomposition is higher (Kowalczyk, 1978), and is lowest for highly humified moorsh (Table 2.11). The CO_2 evolution was shown to be maximum at groundwater level of 90 cm (Table 2.12). Lysimeter studies in drained organic soils showed losses from 2.8 to 6.7 Mg CO_2 ha^{-1} yr^{-1}. Considering contributions of greenhouse gases relative to CO_2, a climatic

Table 2.9   Average Concentration of Organic Substances in Two Peat-moorsh Soil Profiles in Poland

Soil	Horizon	Depth (cm)	Bitumens kg kg⁻¹ dry matter	Hemicellulose	Cellulose	Lignin	HA[a]	FA[a]	HA + FA
						(kg C) (kg total C)⁻¹			
Topola-Blonie	M₁	5–10	0.0404	0.0940	0.0642	0.2520	0.3952	0.1571	0.5523
	M₂	15–20	0.0408	0.0869	0.0528	0.2369	0.4458	0.1290	0.5748
	M₃	23–28	0.0469	0.0711	0.0502	0.2107	0.4811	0.1373	0.6184
	T₁	40–45	0.0501	0.0737	0.0660	0.3835	0.3164	0.0973	0.4137
	T₂	75–80	0.0568	0.0703	0.0619	0.3710	0.2982	0.1094	0.4076
Kuwasy	M₁	5–10	0.0780	0.0579	0.0396	0.1428	0.4943	0.2080	0.7023
Szymany	M₂	15–20	0.0809	0.0657	0.0469	0.1706	0.4960	0.1257	0.6227
	M₃	25–30	0.1003	0.0528	0.0488	0.2197	0.4586	0.1198	0.5784
	T₁	45–50	0.1088	0.0486	0.0439	0.2839	0.4456	0.1034	0.5490
	T₂	90–95	0.0960	0.0327	0.0359	0.2496	0.4920	0.1032	0.5952

[a] HA = humic acids; FA = fulvic acids.

Source: From Okruszko, 1960. Roczniki Nauk Rolniczych, F74:5–89. With permission.

Table 2.10 Composition of Ashes in 14 Peat-Moorsh Soils of Poland

Component	Horizon (g kg⁻¹)				
	M_1	M_2	M_3	T_1	T_2
Ash	329 ± 87	226 ± 53	174 ± 37	135 ± 32	102 ± 27
Silica – SiO_2	167 ± 75	94 ± 53	48 ± 30	32 ± 27	9 ± 4
Potassium – K_2O	0.8 ± 0.21	0.86 ± 0.55	0.47 ± 0.20	0.42 ± 0.25	0.29 ± 0.07
Sodium – Na_2O	1.89 ± 1.34	2.11 ± 7.12	2.54 ± 2.24	3.08 ± 3.08	2.34 ± 2.57
Calcium – CaO	46.5 ± 20.8	42.5 ± 11.7	37.8 ± 9.3	36.4 ± 9.2	32.2 ± 7.5
Magnesium — MgO	3.21 ± 1.2	3.2 ± 2.1	2.7 ± 2.2	2.2 ± 1.6	2.6 ± 3.3
Phosphorus – P_2O_5	4.68 ± 4.0	3.21 ± 1.02	1.90 ± 0.32	1.66 ± 0.44	1.40 ± 0.32
Iron – Fe_2O_3	44.5 ± 26.1	34.8 ± 19.1	33.5 ± 16.3	17.0 ± 8.0	15.6 ± 4.8

Source: From Okruszko, 1960. *Roczniki Nauk Rolniczych,* F74:5–89. With permission.

Table 2.11 Organic Matter and Nitrogen Mineralization in the 0–20 cm Arable Layer as Related to Moorsh Stage

Moorsh Degree	OMC[a] (Mg ha⁻¹)	Mineralized OM[a] (Mg ha⁻¹ yr⁻¹)	Total N (Mg N ha⁻¹)	Mineralized N (N_{min}) (kg N ha⁻¹ yr⁻¹)	(%) (N_{min}/N_{total})
Mtl	275	2.2–2.8	9.60	77–98	0.8–1.0
MtlI	414	6.5–11.4	17.86	281–493	1.6–2.8
MtlII	519	4.0–8.7	21.92	170–369	0.8–1.7
MtlII with a dry, grain moorsh	468	1.8–3.3	26.00	99–186	0.4–0.7

[a] OMC = Organic matter content; OM = Organic matter.

Source: From Frackowiak, 1969. Intensive nitrogen mineralization in moorsh (in Russian with German summary), in *Transformations in Organic Soils under the Influence of Drainage and Reclamation.* Anonymous. Minsk, Belarus, 182–195. With permission.

impact factor (CO_2 + CH_4 + NO_x) was computed to be between 2.9 and 10.3 Mg CO_2 ha⁻¹ yr⁻¹ in drained fen organic soils (Table 2.13).

Soluble substances percolate to groundwater as well as ditches, rivers, and lakes as a result of a positive water balance in drained organic soils. The N leaching depends on soil type, intensity of land use, and groundwater level (Figure 2.2). Some nitrate may reach the groundwater through macropores under intensive drainage. In acid fen soils, the N leaching rate is small (Table 2.14). In calcareous fen soils, N leaching rates of 40–80 kg N ha⁻¹ yr⁻¹ (Scheffer, 1994), sometimes exceeding 140 kg N ha⁻¹ yr⁻¹ with groundwater levels deeper than 2 m (Rück, 1991), have been recorded. Solute transfer generally increases with the advancement of MFP.

V. CHANGES IN BIOLOGICAL PROPERTIES

Heterotrophic bacteria, fungi, amoeba, and microalgae are the dominant groups of microorganisms in European peatlands. The relative importance of heterotrophic organisms (73 to 95% of total biomass) and the predominance of testate amoeba in the protozoan biomass are typical of *Sphagnum*-dominated peatlands (Mitchell et al.,

Table 2.12 Carbon Mineralization and Organic Soil Subsidence as Related to Groundwater Level

Specification	Unit	Groundwater Level			
		30 cm	60 cm	90 cm	120 cm
		Peat Thickness of 50 cm: 57 kg C m^{-2}			
C mineralization	g C m^{-2}	286.00	398.00	490.00	374.00
Soil subsidence	Cm yr^{-1}	0.18	0.25	0.31	0.24
		Peat Thickness of 150 cm: 103 kg C m^{-2}			
C mineralization	g C m^{-2}	391.00	562.00	669.00	658.00
Soil subsidence	Cm yr^{-1}	0.41	0.59	0.70	0.69

Source: From Mundel, 1976. *Archiv für Acker Pflanzenbau und Bodenkunde*, 20:669–679.

Table 2.13 Climatic Impact of Gas Emissions from Wet and Drained Fens in Germany

Greenhouse Gas	Wet Fens		Drained Fens	
	Net Emission (kg C or N ha^{-1} yr^{-1})	Climatic Impact (kg C ha^{-1} yr^{-1})	Net Emission (kg C or N ha^{-1} yr^{-1})	Climatic Impact (kg C ha^{-1} yr^{-1})
CO_2	−140 to −2.250[a]	−140 to −2250	2900 to 6700	2900 to 6700
CH_4	2.7 to 521	24 to 4585	−1.4[a] to 3.3	−12[b] to 29
N_2O	0.0 to 0.8	0 to 107	0.3 to 26.9	40 to 3605
Total impact	—	−2.226[b] to 4.552	—	2928 to 10,334

[a] Negative number means a net sink of gases in peatland ecosystems.
[b] Negative number means that the organic soil ecosystems attenuate the impact of greenhouse gases.

Source: From Augustin, 2001. Emission, trapping and climatic importance of trace gases (in German), in *Landschaftsoekologische Moorkunde*. Succow, M. and Joosten, H., Eds., E. Schweizerbart'sche Verlagsbuchhandlung (Nägele u. Obermiller), Stuttgart, Germany, 28–40. With permission.

1999). Actinomycetales, which are adapted to low moisture, high periodic temperature, and lack of soluble organic compounds (Burzynska-Czekanowska, 1967), prevail (76 to 80% of total number of microbes) in ill-managed moorsh soils (Maciejewska, 1956). They decrease abruptly to 27% in well-managed areas (water control and fertilization), where bacteria increase in number from 12–14% to a maximum of 65% (Maciejewska, 1956). Luthardt (1987) and Behrendt (1995) found that the bacteria feeding activity declined under intensive compared with extensive use, and that soil animals and microbes interactions producing stable humic substances did not occur in moorsh fen soils, thus leading to complete mineralization of organic substances.

Mites and springtails make up 90% of the peat-moorsh mesofauna. Unmanaged grassland soils harbor 12 times more mesofauna than arable soils and four times more than in soils under managed pasture, especially in the layer immediately below surface (0–5 cm). Soil mesofauna was found to be 3 to 4 times less abundant in grassland organic soils slightly affected by MFP compared to similar soils strongly affected by MFP (Gawlik, 1971).

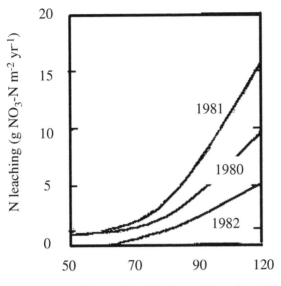

Figure 2.2   Relationship between groundwater level and N-leaching (Drawn from: Behrendt et al., *Zeitschrift für Kulturtechnik und Landentwicklung*, 35:200–208,1993. With permission.)

Table 2.14 Mean Annual Nutrient Loss from a Peatland with 200 mm Runoff Water per Year

Mire Type and Use	N (kg ha⁻¹)	P (kg ha⁻¹)	K (kg ha⁻¹)	Ca (kg ha⁻¹)
Raised mires				
Uncultivated	5	1.3–1.7	10–20	12–20
Grassland	2–20	4–9	20–30	34–45
Arable	10–40	8–17	ND	ND
Fens				
Grassland	5–20	0.1–2.0	10–50	20–150
Arable	40–80	0.1–5.0	20–50	20–150

ND = not determined.

Source: From Scheffer, 1994. *Dynamics of Elements in Fen Soils* (in German). NNA-Berich+ 7 H, 2:67–73, Niedersächsische Naturschutz Akademie, Schneverdingen, Lower Saxonia, Germany. With permission.

Makulec and Chmielewski (1994) examined the species, numbers, and biomass of earthworm communities in response to drainage in the Biebrza fen region of Poland. Mean densities and biomass of earthworms markedly increased during the first few years after drainage. Most sites were dominated by *Lumbricus rubellus*, which accounted for 60 to 80% of the earthworm population and 80 to 90% of its biomass. In a meadow drained several decades before, earthworms deposited 40–100 Mg ha⁻¹ of cast dry matter during the growing season. The number of fungi increased in earthworm casts with the advancement of MFP, but that of *Actinomycetes* decreased. Earthworms reduced the number of microorganisms in soils rich in bacteria, and increased microbial populations in soils poor in microflora.

Wasilewska (1994) studied nematodes as bio-indicators of MFP in alder fens. The critical values of four bioindicators were assessed in relation to a high degree of MFP as follows:

- More than 5×10^6 individuals m^{-2} for *Paratylenchus* sp.,
- Less than 2% abundance for omnivores and predators in a community
- Ratios of the number of bacteria and fungivores to the number of obligatory plant parasites less than 2
- A maturity index less than 1.7

The diversity of seven ant species in natural and drained fen soils was investigated by Petal (1994). Nest density and frequency were lower in soils derived from alder peat compared to sedge or sedge-moss peat materials. Number and abundance of ant species, as well as nest frequency and density, were lower in pristine and previously drained alder peats than in recently drained sites. The soil matrix of ant nests contained more humic substances and exchangeable cations compared with the surrounding soil matrix.

VI. SOIL DEGRADATION SYMPTOMS AND PREVENTION

Advanced MFP in the top soil and structural damage in the subsoil are typical features of degraded soils (Table 2.15). Shallow fen organic soils are highly degraded when a layer with laminar structure hinders water infiltration at a depth of 15–25 cm below soil surface (Figure 2.3). Such a layer forms a physical barrier to root penetration and capillary rise. The soil surface becomes dry and resistant to rewetting and penetration by water. Especially in the spring, the bare soil is highly susceptible to wind erosion. Schäfer et al. (1991) reported soil losses of 3 to 7 Mg ha^{-1} d^{-1} with a wind speed of 7 to 13 m s^{-1}.

The MFP is alleviated by grassland farming and groundwater level maintained close to 30 cm below the soil surface. Mineralization and gas emissions increase markedly when the water level is drawn below 30 cm. Maximum forage yield is obtained at a groundwater level between 40 and 60 cm (Wessolek, 1999). More intensive soil management practices requiring deeper water table levels are thus at increasing risk to the environment.

VII. CONCLUSION

In drained organic soils, the moorsh-forming process is characterized by repeated shrinkage and swelling, as well as loss of organic substances in upper soil layers. Indicators of change in physical properties are soil morphology, volume of the solid phase, soil porosity and water regime. Chemical indicators are the composition of organic and inorganic materials, N and C mineralization rates, and the quality of drainage water. Bioindicators were proposed in Europe for microorganisms and the mesofauna. A minimum dataset for assessing soil degradation depends on requirements for sustainable management of soil–plant systems.

Table 2.15 Trend in Soil Attributes and Functions with Advancement of the Moorsh-forming Process

Increase	Decrease
Attribute	
Bulk density (0.1 → 0.4 g cm^{-3})	Total porosity: > 0.90 → < 0.80 m^3 m^{-3}
Ash content (0.10 → 0.70 kg kg^{-1})	C/N ratio: 30 → 10, dimensionless
Permanent wilting point (0.10 → 0.30 m^3 m^{-3})	Macro- and mesopores: 0.25 → < 0.10 m^3 m^{-3}
Cation exchange capacity: 50 → 450 mmol L^{-1}	Field capacity: > 0.50 → < 0.30 m^3 m^{-3}
	Permeability (kf): > 100 → < 10 cm d^{-1}
	Unsaturated hydraulic conductivity (ku): > 5 → 0.5 mm d^{-1}
Function	
1. Medium for plant growth	
Nutrient requirements	
Temperature variation	
Irregularity of micro relief	
2. Regulation and partitioning of water flow	Groundwater level
Risk of water-logging	
Risk of drought	
Risk of wind erosion	
3. Environmental buffer for hazardous compounds	Denitrification potential
Nitrification: 0 → 1000 kg NO$_3$ N ha^{-1} yr^{-1}	Emission of methane
Total emission of greenhouse gases	

Source: From Zeitz, 1991. *Zeitschrift für Kulturtechnik und Landentwicklung, 32:* 227–234; Luthardt, 1993. *Naturschutz und Landschaftspflege in Brandenburg, Sonderheft Niedermoore,* 35–40; Roeschmann et al., 1993. *Geologisches Jahrbuch,* F29: 3–49. With permission.

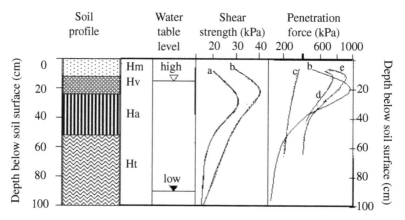

Figure 2.3   Barrier layer in a degraded organic soil: a = MtII Putzar soil, b = MtIII Klein-Markow soil; c = fen peat; d = MtI fen organic soil; e = MtII fen organic soil. (Adapted from Zeitz et al., *Feldwirtschaft*, 28:211–216, 1987. With permission.)

REFERENCES

Augustin J. 2001. Emission, trapping and climatic importance of trace gases (in German), in *Landschaftsoekologische Moorkunde.* Succow, M. and Joosten, H., Eds., E. Schweizerbart'sche Verlagsbuchhandlung (Nägele u. Obermiller), Stuttgart, Germany, 28–40.

Baden, W. and Eggelsmann, R. 1963. Permeability of organic soils (in German). *Zeitschrift für Kulturtechnik und Flurbereinigung,* 4:226–254.

Behrendt, A. 1995. Peat wastage in northern fen mires of Germany, a contribution to mire conservation (in German). Dissertation, Landwirtschaftlich Gärtnische Fakultät der Humboldt Universität zu Berlin, 179 pp.

Behrendt, A., Mundel, G., and Hölzel, D. 1993. Studies on carbon and nitrogen transformations in fen peat soils using lysimeters (in German). *Zeitschrift für Kulturtechnik und Landentwicklung,* 35:200–208.

Burghardt, W. and Ilnicki, P. 1978. Subsidence in repeatedly drained bog soils in flatlands of northwestern Germany. Report 4: Change in soil water content and pore-size distribution as a result of subsidence (in German). *Zeitschrift für Kulturtechnik und Flurbereinigung,* 19:146–157.

Burzynska-Czekanowska, E. 1967. The flora of Actinomycetes in important types of peat (in Polish with English summary). *Zeszyty Problemowe Postepow Nauk Rolniczych,* 76:313–353.

Frackowiak, H. 1969. Intensive nitrogen mineralization in moorsh (in Russian with German summary), in *Transformations in Organic Soils under the Influence of Drainage and Reclamation.* Anonymous. Minsk, Belarus, 182–195.

Gawlik, J. 1971. Effect of use of moorsh soils on moisture-air conditions and soil fauna (in Polish with English summary). *Annales Universitatis Mariae Curie-Sklodowska,* E 26:375–395.

Gotkiewicz, J., Kowalczyk, Z., and Okruszko, H. 1975. Mineralization of nitrogen and carbon compounds in peat-moorsh soils under different air and moisture conditions (in Polish with English summary). *Roczniki Nauk Rolniczych,* F79.1:131–150.

Gotkiewicz, J. and Szuniewicz, J. 1987. Transformation of sites and soils on agricultural experimental areas (in Polish with English summary). *Bibl. Wiadomosci Instytutu Melioracji i Uzytkow Zielonych,* 68:33–41.

Ilnicki, P. 1967. Peat shrinkage by drying as related to structure and physical properties (in Polish with English summary). *Zeszyty Problemowe Postepow Nauk Rolniczych,* 76:197–311.

Ilnicki, P. 1982. Hysteresis of the water tension curve in organic soils (in German with English summary). *Zeitschrift für Pflanzenernährung und Bodenkunde,* 145:363–374.

Kowalczyk, Z. 1978. Effect of moisture content in peat-muck soils on biological activity as determined by the cellulose test (in Polish with English summary). *Roczniki Nauk Rolniczych,* F79:67–78.

Luthardt, V. 1987. Ecological studies on deep fens at different developmental stage and used in agriculture (in German). Dissertation, Akademie der Landwirtschaftwissenschaften der DDR, Berlin, Germany, 183 pp.

Luthardt, V. 1993. Development objectives for fen regions with an example from Sernitz-Niederung near Greiffenberg (in German). *Naturschutz und Landschaftspflege in Brandenburg, Sonderheft Niedermoore,* 35–40.

Maciejewska, Z. 1956. The interactions between peat soil microflora, land use and plant cover (in Polish). *Zjazd Naukowy Polskiego Towarzystwa Gleboznawczego,* IX:20–23.

Makulec, G. and Chmielewski, K. 1994. Earthworm communities and their role in hydrogenic soils, in *Proc. Int. Peat Soc. Symp.,* Jankowska-Huflejt, H. and Golubiewska, E., Eds., Warsaw-Biebrza, Poland, 417–427

Mitchell, E.A.D. et al. 1999. The microbial loop at the surface of five *Sphagnum*-dominated peatlands in Europe: Structure and effects of elevated CO_2. *Proc. Int. Peat Soc. Symp.,* Finnish Peatland Society, Jokioinen, Finland, 86.

Mundel, G., 1976. Mineralization of fen peat (in German). *Archiv für Acker Pflanzenbau und Bodenkunde,* 20:669–679.

Okruszko, H. 1960. Muck soils of valley peatlands and their chemical and physical properties (in Polish with English summary). *Roczniki Nauk Rolniczych,* F74:5–89.

Okruszko, H. 1971. Determination of specific gravity of hydrogenic soils on the basis of their mineral particles content (in Polish with English summary). *Wiadomosci Instytutu Melioracji i Uzytkow Zielonych,* 10.1:47–54.

Okruszko, H. 1976. Keys to hydrogenic soil investigation and classification for reclamation purposes (in Polish with English summary). *Bibl. Wiadomosci Instytutu Melioracji i Uzytkow Zielonych,* 52:7–54.

Okruszko, H. 1985. Decession in the natural evolution of low peatlands. *Intecol. Bull.,* 12:89–94.

Okruszko, H. 1993. Transformation of fen-peat soil under the impact of draining. *Pol. Akad. Nauk,* 406:3–75.

Okruszko, H. 1994. System of hydrogenic soil classification used in Poland. *Bibl. Wiadomosci Instytutu Melioracji i Uzytkow Zielonych,* 84:5–27.

Petal, J. 1994. Ant communities in Biebrza marshland, in *Proc. Int. Peat Soc. Symp.* Jankowska-Huflejt, H. and Golubiewska, E., Eds., Warsaw-Biebrza, Poland, 399–406.

Roeschmann, G. et al. 1993. Mires in soil classification (in German). *Geologisches Jahrbuch,* F29:3–49.

Rück, F. 1991. Nitrogen content and nitrate leaching in calcareous fen soils under arable farming and abandoned land (in German), in VDLUFA Soil Excursion in Ulm, Verband Deutscher Landwirtschaftlicher Untersuchungs und Forschungsanstalten, Darmstadt, Germany, 51–55.

Sauerbrey, R. and Zeitz, J. 1999. Peatlands. Section 3.3.3.7, in *Handbuch der Bodenkunde.* (in German). Blume, H.P., Ed., Loseblätter Ausgabe, Ecomed Publ., Landsberg, Germany, 20 pp.

Schäfer, W., Neemann, W., and Kuntze, H. 1991. Wind erosion on calcareous fenlands in North Germany. *Fenland Symp.*, Cambridge, England.

Schäfer, W. 1996. Changes in physical properties of organic soils induced by land use. *Proc. 10th Int. Peat Congr.*, 4:77–84.

Scheffer, B. 1994. Dynamics of elements in fen soils (in German). NNA-Berich+ 7 H, 2:67–73, Niedersächsische Naturschutz Akademie, Schneverdingen, Lower Saxonia, Germany.

Schmidt, W. and Illner, K. 1976. Soil formation in fen soils used in agriculture (in German). *Melioration Landwirtschaftsbau*, 10:166–168.

Sponagel, H. et al. 1996. *Methods of Soil Cartography* (in German). E. Schweizerbart'sche Verlagsbuchhandlung (Nägele u. Obermiller), Stuttgart, Germany, 392 pp.

Szuniewicz, J. and Szymanowski, M. 1977. Physico-hydrological properties and the formation of air-water conditions on distinguished sites of the Wizna fen. *Polish Ecol. Stud.*, 3(3):17–31.

Titze, E. 1966. Water and air composition of the upper earthy layer of the Klenzer fen and its influence on yield (in German). Dissertation, Universität Rostock, 112 pp.

TGL 31222/03. 1985. Standard methods in soil physics — Density and porosity (in German). Staatlich Technische Normen, Gütevorschriften und Lieferbedingungen Standards der DDR, Berlin.

Wasilewska, L. 1994. Changes in biotic indices in meadow ecosystems after drainage of fens, in *Proc. Int. Peat Soc. Symp.* Jankowska-Huflejt, H. and Golubiewska E., Eds., Warsaw-Biebrza, Poland, 407–416

Wessolek, G. 1999. Processes of soil formation in fen mires (in German). *Ökologische Hefte, Landwirtschaftlich Gärtnische Fakultät der Humboldt Universität zu Berlin*, 11:96–122.

Zeitz, J. 1991. Determination of the permeability of fen soils across soil formation stages (in German). *Zeitschrift für Kulturtechnik und Landentwicklung*, 32:227–234.

Zeitz, J. 1992. Physical properties of soil horizons in fen peat soils used in agriculture (in German). *Zeitschrift für Kulturtechnik und Landentwicklung*, 33:301–307.

Zeitz, J., Titze, E., and Kosov, W. 1987. Influence of excessive drainage on soil condition and yield in deep fen peat (in German). *Feldwirtschaft*, 28:211–216.

Zeitz, J. and Tölle, R. 1996. Soil properties of genetically originated peat horizons in drained fens. *Proc. 10th Int. Peat Congr.*, 2:198–206.

CHAPTER **3**

Water-Related Physical Attributes
of Organic Soils

Tomasz Brandyk, Jan Szatylowicz, Ryszard Oleszczuk, and Tomasz Gnatowski

CONTENTS

ABSTRACT

    Conservation of organic soils used for agriculture requires proper regulation and
partitioning of water flow in the environment. This chapter reviews the physical

properties involved in water retention and transfer in drained organic soils. Basic physical properties are bulk density, specific density, porosity, and ash content. Water retention characteristics of organic soils are influenced by the degree of peat decomposition. Water retention characteristics can be derived from other soil properties using pedotransfer functions. The results of saturated and unsaturated hydraulic conductivity measurements must be interpreted in reference to factors influencing their determination. Relationships between concomitant changes in soil moisture and volume during shrinkage are illustrated by characteristic curves. For converting volume changes into crack volume and subsidence, a dimensionless shrinkage geometry factor can be used. The authors investigated:

1. The influence of moisture content on water repellency
2. The effect of repellency on field moisture distribution patterns
3. The spatial variability of bulk density, hydraulic conductivity, moisture retention characteristics, and moisture content at plot scale

Shrinkage characteristics, the shrinkage geometry factor, and parameters describing water repellency should be incorporated into hydrological models examining simultaneously water transport and the subsidence of organic soils.

I. INTRODUCTION

Peatland hydrology is fundamental to understanding, quantifying, and evaluating the key soil function of water regulation and partitioning in the environment. Water is the driving force for peatland formation (carbon sink) and maintenance (biodiversity and productivity). This is why peatland hydrology has become a research area of high priority worldwide. To evaluate hydrological phenomena, such as water storage, water table fluctuation, and evapotranspiration in peatlands, water-related physical properties of organic soils must be quantified. A peculiar property of organic soils is that they originate *in situ* and undergo transformations in response to changes in water conditions.

The main purpose of this chapter is to review soil properties important for water retention and conduction in organic soils. Special attention is also given to the shrinkage process, water repellency, and spatial variability of organic soil properties.

II. BASIC PHYSICAL PROPERTIES

Soil consists of solid, liquid, and gaseous phases. The solid phase of organic soils is made of plant fibers, humus, and mineral matter such as grains of different sizes (from sand to clay) as well as amorphous substances in the form of carbonates, phosphates, and hydroxides. The rate at which plant materials in mires are decomposed depends on many factors such as acidity, temperature, moisture, oxygen supply, biochemical makeup, as well as peat organisms in terms of composition and number. The most widely used method to determine the degree of peat decomposition is the Von Post method (Von Post, 1922) with its 10 classes of humification (i.e., H1 referring to undecomposed peat and H10 to completely decomposed peat).

Soil bulk density is soil mass per unit volume. Peat bulk density is determined by dividing the oven dry (105°C) peat mass by the volume of a core of undisturbed peat samples. The bulk density of peat deposits varies according to botanical composition and degree of peat decomposition. Moss peat generally shows a smaller bulk density than fen peat, mainly due to a lower degree of decomposition and smaller ash content. With an increasing degree of decomposition, an increase in bulk density is often observed. Päivänen (1973) reported a positive and approximately linear relationship between bulk density and the Von Post humification scale for *Sphagnum* and *Carex* peats. Increase in bulk density with increasing degree of decomposition was smallest with *Sphagnum* peat and largest with sedge peat. In peat deposits, bulk density increases with depth, primarily due to the burden of overlying peat layers. Bulk density values of Finnish peats from undisturbed and drained areas varied from 0.04 to 0.20 g cm^{-3} (Päivänen, 1973). Values for Minnesota peats ranged from 0.02 to 0.26 g cm^{-3} (Boelter, 1969). Values as high as 0.2 to 0.4 g cm^{-3} have been reported for fen peats of Central Europe (Okruszko, 1993).

Particle density is the dry mass of solids divided by solid volume. Average particle density of the organic soil mass is 1.45 g cm^{-3}, varying slightly from 1.3 to 1.6 g cm^{-3}, depending on degree of decomposition (Okruszko, 1971). Such variations in particle density are small compared with bulk density. For peat materials in an advanced stage of decomposition, particle gravity is greatest for woody peat. Peat particle density depends largely on ash content.

Ash content is determined by igniting dried peat in a muffle furnace at about 550°C until constant weight. Ash content is expressed as the percentage of ignited residue to the quantity of dry matter. Ash content of sedge and woody peats is considerably higher than that of *Sphagnum* peat. In general, ash content is higher in fen peat than in bog peat. Okruszko (1971) obtained a linear relationship between particle density (ρ_p in g cm^{-3}) and ash content or loss on ignition (M in %) from 2996 peat samples containing 0.7 to 99.5% ash (7 to 995 g kg^{-1}), as follows:

$$\rho_p = 0.011M + 1.451, \quad r = 0.96 \qquad (3.1)$$

Peat particle density thus increases by 0.011 g cm^{-3} for each 1% or 10 g kg^{-1} increase in ash content above 1.451 g cm^{-3}.

Peat is a highly porous material. The pores differ in size and shape, depending on the geometry of plant residues and on degree of peat decomposition. Total porosity can be assessed from bulk density and particle density. Total porosity of peat is about 0.97 m^3 m^{-3} for undecomposed peats and 0.81–0.85 m^3 m^{-3} for highly decomposed peats (Boelter, 1969; Päivänen, 1973). Peat is also characterized by its volume percentage of the solid phase, computed as the ratio of bulk density to particle density. According to Okruszko (1993), the mean volume of the solid phase is 0.08 m^3 m^{-3} for slightly decomposed peat, 0.10 m^3 m^{-3} for moderately decomposed peat, and 0.11 m^3 m^{-3} for highly decomposed peat.

III. SOIL WATER CHARACTERISTIC

A. Methods of Determination

The water retention curve refers to the relationship between soil water content and matric potential. The exact relationship can be determined in the laboratory using undisturbed soil samples, a sand table, and pressure chambers (Klute, 1986), or directly in the field using tensiometers and time-domain reflectometers (TDR). From this curve, we obtain plant-available water, defined as the amount of water held by a soil between field capacity and wilting point. In the literature, soil water matric potential at field capacity ranges from about 50 to 500 cm (pF 0.7 to 2.7 or −5 to −50 kPa). Water content at pF 4.2 has been usually considered as the permanent wilting point or the lower limit of plant-available water.

Determining the water retention curve directly can be expensive and time consuming. For mineral soils, many attempts have been made from easily measured standard soil properties (Tietje and Tapkenhinrichs, 1993). Pedotransfer functions for predicting the water retention curve can be divided into two main types: point estimation and parametric estimation. Point estimation is an empirical function that predicts water content at a predefined potential. It provides data in a tabular form, which complicates mathematical and statistical operations. Parametric estimation of pedotransfer functions is based on the assumption that the relationship between water content (θ) and matric potential (h) can be described adequately by a hydraulic model (e.g., Van Genuchten, 1980). Empirical functions were developed to estimate parameters of the hydraulic model from easily measured properties. It yields a continuous function for $\theta(h)$, thus facilitating mathematical and statistical operations.

For organic soils, very few attempts have been made to estimate water content at a predefined potential from certain peat properties. Boelter (1969), Päivänen (1973) and Szymanowski (1993a) used regression equations to relate water content at certain pressure head values to peat bulk density. Boelter (1969) developed empirical equations for Minnesota moss and herbaceous peats, as well as peats with a high wood content. Bulk density of peat samples ranged from 0.02 to 0.25 g cm^{-3}. Equations developed by Boelter (1969) are listed in Table 3.1. In Finland, Päivänen (1973) studied *Sphagnum*-dominated peat materials varying in degree of humifica-

Table 3.1   Regression Equations Relating Volumetric Moisture Content (θ in %) to Bulk Density (ρ_b in g cm^{-3}) at Different Soil Water Matric Potentials

Matric Potential (pF)	Regression Equations	R²
0.7	$\theta = 39.67 + 638.29\rho_b - 2010.89\rho_b^2$ ‡	0.70
2.0	$\theta = 2.06 + 719.35\rho_b - 1809.68\rho_b^2$	0.88
4.2	$\theta = 1.57 + 115.28\rho_b - 107.77\rho_b^2$	0.82

Source: From Boelter, D.H. 1969. *Soil Sci. Soc. Am. Proc.*, 33:606–609. With permission.

Table 3.2   Regression Equations Relating Volumetric Moisture Content (θ in %) to Bulk Density (ρ_b in g cm^{-3}) in Fen Peat at Different Soil Water Matric Potentials

Matric Potential (pF)	Type of Peat	Regression Equation	No. of Samples	r²
2.7	Sedge moss	$\theta = 18.43 + 189.46\rho_b$	232	0.576
	Tall sedge	$\theta = 28.98 + 143.90\rho_b$	264	0.295
	Reed	$\theta = 38.63 + 94.19\rho_b$	207	0.237
	Alder	$\theta = 48.22 + 65.35\rho_b$	201	0.170
	Moorsh	$\theta = 40.52 + 38.10\rho_b$	552	0.045
4.2	Sedge moss	$\theta = -4.43 + 160.29\rho_b$	232	0.773
	Tall sedge	$\theta = 2.10 + 128.16\rho_b$	264	0.456
	Reed	$\theta = 2.38 + 128.76\rho_b$	207	0.527
	Alder	$\theta = 8.07 + 97.29\rho_b$	201	0.404
	Moorsh	$\theta = 10.75 + 66.66\rho_b$	552	0.441

Source: From Szymanowski, M. 1993a. *Wiadomosci Instytutu Melioracji i Uzytkow Zielonych*, XVII(3):153–174. With permission.

tion. Bulk density was in the range between 0.037 and 0.207 g cm^{-3} and regression equations were similar. Szymanowski (1993a) analyzed 1588 fen peat samples from the Biebrza River Valley and proposed empirical regression equations relating bulk density to moisture contents at pF values of 2.7 and 4.2. Bulk density ranged from 0.136 g cm^{-3} for moss peat up to 0.233 g cm^{-3} for moorsh layers. The regression equations developed by Szymanowski (1993a) are presented in Table 3.2.

Weiss et al. (1998) tested continuous moisture retention models for organic soils and found that the Van Genuchten's model (1980) was most suitable if residual water content was omitted. The model was presented in the following form:

$$\theta = \theta_s[1 + (\alpha h)^n]^{-1+1/n} \tag{3.2}$$

where θ is moisture content (m^3 m^{-3}), θ_s is saturated moisture content (m^3 m^{-3}), $\mathbf{h}$ is pressure head in cm H$_2$O, α (cm^{-1}) and $\mathbf{n}$ (dimensionless) are parameters defining the Van Genuchten curve shape. Weiss et al. (1998) proposed to evaluate shape parameters required in Equation 3.2 as follows:

$$n = 1.49 - 3.56\rho + 13.2\rho^2 + 0.00027C - 0.037\text{Layer1}$$
$$\log_{10}(\alpha) = -51.8\rho^2 - 0.0057C - 0.01S + 0.53\text{Layer1} \tag{3.3}$$

where ρ is bulk density (g cm^{-3}), C and S are *Carex* and *Sphagnum* content as percentages, respectively, **Layer 1** is the layer located 0–10 cm below peat surface and is a qualitative variable having a value of 1 or 0. Moisture content at saturation (θ_s) was obtained from sample porosity. The percentage of botanical components was included in Equation 3.3 because the difference in water retention between different peat types can be explained not only by differences in peat characteristics related to bulk density, but also by differences in plant residues, cell structure, and peat pore geometry. The pedotransfer functions developed by Weiss et al. (1998) were based on 152 peat samples collected from 38 undrained and drained pine mires in Finland with soil bulk density ranging from 0.04 g cm^{-3} to 0.18 g cm^{-3}. Pedo-transfer functions for predicting the water retention curve of organic soils were also developed by Wösten et al. (1999), who used the Food and Agriculture Organization (FAO) definition of histic horizons. Moisture retention curves used in the previously described models were based on desorption curves, and hysteresis effects were not taken into account. Effects of swelling and shrinkage were also neglected.

B. Water Retention of Peat and Moorsh Materials

The relationship between matric potential and soil moisture content in organic soils depends on degree of decomposition and botanical composition of peat resi-dues. Soil water characteristics of slightly, partially, or highly decomposed *Sphag-num* peat (high bog peat) and reed peat (fen peat) materials corresponding to the Von Post humification scale of H1–2, H5–6, and H9–10, respectively, are presented in Figure 3.1. Curves for *Sphagnum* peat were derived from measurements performed by Päivänen (1973). The curve for slightly decomposed *Sphagnum* peat (Figure 3.1a) shows a loss of more than 54% of its volumetric moisture content at a matric potential of 50 cm (pF 1.7), while the curve for slightly decomposed reed peat (Figure 3.1b) shows a loss of about 15% of moisture content. The partially or highly decomposed *Sphagnum* peat lost 11–19% of its moisture content at the same matric potential, whereas partially or highly decomposed reed peat lost 4–6%. The content in plant-available water, computed by difference in moisture content between pF values of 1.7 and 4.2, increased from 0.25 to 0.55 m^3 m^{-3} in *Sphagnum* peat and decreased from 0.65 to 0.50 m^3 m^{-3} in reed peat materials with increasing peat decomposition.

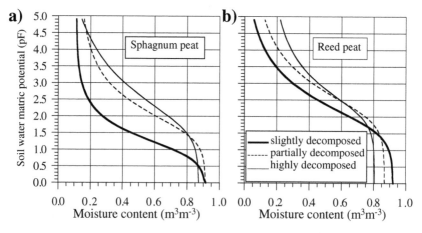

Figure 3.1   Water retention curves for *Sphagnum* (a) and reed (b) peats by degree of decom-position. (Data for *Sphagnum* peats from Päivänen, J. 1973. *Acta For. Fenn.*, 129:1–70. With permission.)

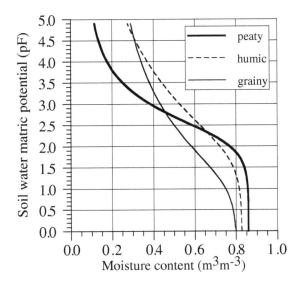

Figure 3.2 Water retention curves of moorsh materials.

In both cases, an increase in degree of decomposition was associated with an increase in moisture content at wilting point.

Drainage and intensive use of peatlands lead to the moorsh-forming process (MFP) (Okruszko, 1976). The *moorshing* of organic soils comprises biological, chemical, and physical changes driven by a decrease in water content and an increase in air content. A moorsh is formed in the top layers. The basic feature differentiating the moorsh from the peat layers is soil structure: the moorsh is usually grainy, while the peat ranges from fibrous to amorphous depending on the degree of humification. Okruszko (1976, 1993) divided moorsh formations into three types:

1. The peaty moorsh has plant residues that are macroscopically visible.
2. The humic moorsh shows a crumbly structure.
3. The grainy moorsh has a grainy structure with frequent hard grains formed by humus condensation.

The amount of plant-available water may decline from 0.67 m^3 m^{-3} in peaty moorsh to 0.31 m^3 m^{-3} in grainy moorsh (Figure 3.2). The moisture content corresponding to the permanent wilting point rises with the advancement of MFP. The MFP decreases total porosity up to 0.09% in the moorsh compared with the original peat material (Okruszko, 1993).

IV. SATURATED HYDRAULIC CONDUCTIVITY

A. Methods of Determination

Hydraulic conductivity controls infiltration rate through soil surface as well as capillary flux from the groundwater table. Hydraulic conductivity of saturated soils may be measured either in the laboratory (Klute and Dirksen, 1986) or in the field (Amoozegar and Warrick, 1986). Laboratory methods establish rectilinear flow through a sample, and to control not only temperature or solute and gas content in soil water, but also boundary conditions. Field methods minimize loss of structure or change in soil porosity using much larger samples. Hydraulic conductivity of saturated peat layers located below the groundwater table has been measured using the piezometer method and the auger hole method (Rycroft et al., 1975a). The auger hole method assesses saturated hydraulic conductivity in the horizontal direction. The piezometer method can be used to determine the hydraulic conductivity in the vertical direction. The auger hole method gives the average peat hydraulic conductivity between groundwater level and the bottom of the hole.

Hydraulic conductivity is defined by Darcy's law. According to Ingram et al. (1974) and Rycroft et al. (1975b), peat behavior, especially highly decomposed peat, may depart substantially from Darcy's law. The "non-Darcian" behavior was attributed to elastic properties of peat under compression and to the effective stress principle (Hemond and Goldman, 1985). Nevertheless, Hemond and Goldman (1985) argued that Darcy's law remained an appropriate tool for use in wetland hydrological modeling.

In organic soils, a decrease in hydraulic conductivity values with time of measurement was observed by Ivanov (1953) and Bondarenko et al. (1975). Changes with time of hydraulic conductivity in moderately decomposed fen organic soil materials are illustrated in Figure 3.3. After 650 h of laboratory measurements, saturated hydraulic conductivity declined to 70% of its initial value. Such variation

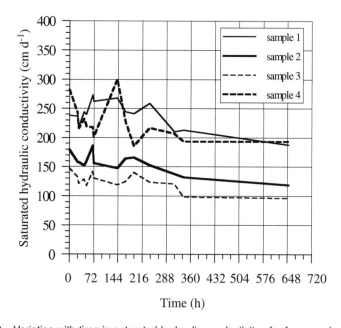

Figure 3.3 Variation with time in saturated hydraulic conductivity of a fen organic soil.

in hydraulic conductivity was explained by the swelling of peat colloids as well as by peat particle migration induced by fluid flow (Ivanov, 1953). Both processes lead to pore blocking, thus reducing local porosity. Bondarenko et al. (1975) found that the decrease in hydraulic conductivity was connected with a change in pore space geometry due to colmatation of soil pores by gas bubbles and other by-products of organic matter decomposition through anaerobic microbiological processes.

Boelter (1965) and Päivänen (1973) found that laboratory evaluation of peat hydraulic conductivity yielded higher values than field evaluation, probably caused by nonconstant flow due to leakage and soil disturbance. Chason and Siegel (1986) also found a general trend for laboratory data to show larger ranges than field data. The smaller ranges of field values may reflect measurements in much larger and thus more representative samples. Laboratory tests may be more affected by peat heterogeneity within the column at a smaller scale.

B. Anisotropy

Peat layers are commonly anisotropic, therefore, hydraulic conductivity is different in the vertical than in the horizontal direction. Ostromecki (1936) found that vertical hydraulic conductivity values (K_v) of fen peat were on average two times larger than horizontal hydraulic conductivity values (K_h), and that the K_v/K_h ratio depended on degree of decomposition. For slightly decomposed fen peat materials, the ratio was greater than 2; for highly decomposed materials, it was equal to 1. Lundin (1964) found higher values for vertical than horizontal hydraulic conductivity in Belorussian fen peat; the K_v/K_h ratio was highest in reed peat and lowest in alder peat. Boelter (1965) found no significant difference between horizontal (measured by the piezometer method) and vertical (measured by the tube method) hydraulic conductivities in slightly to highly decomposed peat materials. Korpijaakko and Radforth (1972) found horizontal saturated hydraulic conductivity value to be greater than the vertical one only close to the surface of a high bog soil. Chason and Siegel (1986) reported that the K_v/K_h ratio was highly variable across peat columns, but that K_h was generally one to two orders of magnitude greater than K_v. They explained their results by the stratification of *Sphagnum* peat. When *Sphagnum* was alive, stem orientation was mainly vertical, thus creating vertical passageways for water. After the plants died, the stems fell over and the decaying process began, thus creating more horizontal planar passageways for water.

The decrease in hydraulic conductivity with increasing depth of *Sphagnum* peat has been observed by Ivanov (1953). This phenomenon was attributed to the acrotelm or "active layer" usually present at the surface of developing mires, characterized by a very loose, open, and porous structure associated with high hydraulic conductivity values. Päivänen (1973) also observed a decrease in hydraulic conductivity with increasing depth in forested Finnish peatlands. Preferential water flow may also be induced by channels resulting from decaying rhizosphere roots or activities of soil invertebrates. In fen peat materials, marked dependence of saturated hydraulic conductivity on depth occurred (Figure 3.4). The values for drained fen organic soils were generally lower compared with undrained soils.

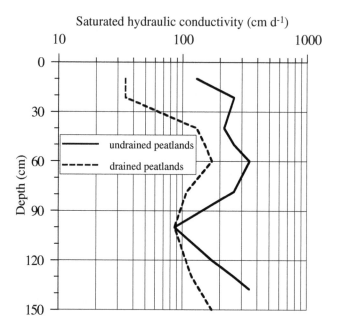

Figure 3.4 Variation in saturated hydraulic conductivity in the profile of a fen organic soil. (Based on data from Lundin, K.P. 1964. *Water Properties of Peat Deposits* (in Russian). Urozaj Press, Minsk, Belarus.)

Hydraulic conductivity of peat deposits varies with degree of peat decomposition (Rycroft et al., 1975a, 1975b). Slightly decomposed peat shows values of the order of 10^{-3} to 10^{-5} m s^{-1}, compared with 10^{-8} m s^{-1} for highly decomposed peat. A negative hyperbolic relationship between saturated hydraulic conductivity and degree of decomposition was obtained by Baden and Eggelsmann (1963) for *Sphagnum*, *Carex* or *Phragmites* peat deposits. The relationship was more pronounced for *Sphagnum* than for *Phragmites* or *Carex* peat deposits. In *Sphagnum* peat, saturated hydraulic conductivity was considerably lower compared with fen peat. In laboratory and field experiments by Korpijaakko (1988), hydraulic conductivity of *Carex* peat was not correlated with degree of decomposition because the structure of the *Carex* peat was already dense even at low degree of decomposition. Table 3.3 presents some field measured values of saturated hydraulic conductivity. Because laboratory methods to determine decomposition are time-consuming and not recommended for routine application, many authors investigated the simple relationship between hydraulic conductivity and easily measured physical properties such as bulk density and volume of solids (Figure 3.5).

C. Seasonal Variation

Seasonal variations of saturated hydraulic conductivity are often observed in swelling clay soils. Water flow through a clay soil is influenced by structural and porosity changes caused by swelling–shrinkage and freezing–thawing cycles during early spring (Messing and Jarvis, 1990). Similar processes were observed by Ole-

Table 3.3 Values of Field-Measured Saturated Hydraulic Conductivity of Organic Soils

Source	Method	Material and Location	Saturated Hydraulic Conductivity (cm d⁻¹)
Lundin (1964)	Auger hole	High bog, Belorussia, H1	250.0
		High bog, Belorussia, H3	50.0
		High bog, Belorussia, H5	5.0
		Sedge-reed fen peat, Belorussia, H3	150.0
		Woody-reed fen peat, Belorussia, H3	1500.0
		Reed fen with channel roots, Belorussia, H3	6000.0
Boelter (1965)	Seepage tube	Slightly decomposed *Sphagnum* peat, Minnesota	3456.0
		Moderately decomposed herbaceous, Minnesota	0.648
Päivänen (1973)	Seepage tube	*Sphagnum* peat, Finland, H1	190.08
		Sedge peat, Finland, H3	181.44
		Sphagnum peat, Finland, H10	0.864
		Sedge peat, Finland, H8	7.344
Lishtvan et al. (1989)	Information not available	*Sphagnum fuscum*, Belorussia, H2	0.0127
		Sphagnum magellanicum, Belorussia, H2	0.0064
Brandyk et al. (1996)	Auger hole	Moderately decomposed sedge-reed fen, Poland	104.0

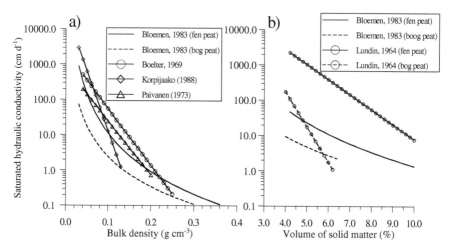

Figure 3.5   Relationships between (a) bulk density and saturated hydraulic conductivity, and (b) volume of solid matter and saturated hydraulic conductivity in organic soils.

szczuk et al. (1995) in organic soils. Seasonal changes occurred in field-measured hydraulic conductivity values using the auger hole method along a 150 m transect in a fen organic soil in summer and autumn (Figure 3.6). Smallest values of saturated hydraulic conductivity were observed in autumn when the soil was swollen, and highest values were shown during summer when soil shrinkage took place.

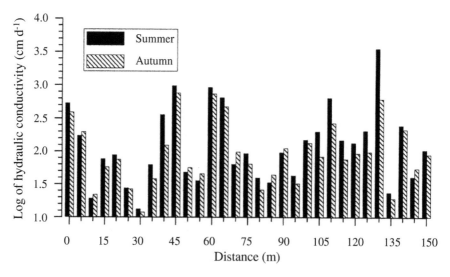

Figure 3.6   Comparison of saturated hydraulic conductivity values in a fen peat measured in summer and autumn along a transect. (From Oleszczuk, R., Szatylowicz, J., and Brandyk, T. 1995. *Przeglad Naukowy Wydzialu Melioracji i Inzynierii Srodowiska*, 7:11–20. With permission.)

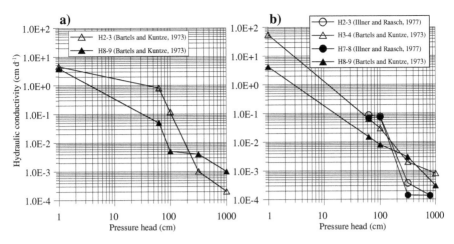

Figure 3.7   Unsaturated hydraulic conductivity of (a) high bog and (b) fen peats with different degrees of decomposition as a function of matric potential.

V. UNSATURATED HYDRAULIC CONDUCTIVITY

Unsaturated hydraulic conductivity is the ability of a porous medium to transmit water when the cross section of the pores is not totally filled with water. To separate this term from the saturated condition, unsaturated hydraulic conductivity is often called *capillary conductivity*. Hydraulic conductivity of unsaturated soils depends on moisture content and matric potential. Many methods have been reported in the literature for the determination of unsaturated hydraulic conductivity (Dirksen, 1991). No universal method is best suited to measure this parameter.

Richards and Wilson (1936) were the first to measure unsaturated hydraulic conductivity as a function of matric potential in pristine and cultivated organic soils. Organic soils showed unsaturated conductivities greater than those observed in mineral soils at low pressure heads, and reached zero values at lower pressure heads compared to mineral soils. Values were presented by Rijtema (1969), Wind (1969), Bartels and Kuntze (1973), Renger et al. (1976), and Illner and Raasch (1977). Unsaturated hydraulic conductivity values determined by Bartels and Kuntze (1973) and Illner and Raasch (1977) using a double-membrane apparatus (Kramer and Meyer, 1968) indicated negative slopes for log–log relationships between unsaturated hydraulic conductivity and pressure head, which depended on degree of peat decomposition (Figure 3.7). Renger et al. (1976) obtained similar results with 11 organic soils that varied in degree of peat decomposition and volume of solids. Bloemen (1983) obtained linear relationships between slopes of unsaturated hydraulic conductivity functions and both bulk density and volume of solids (Figure 3.8). For the same volume of solids, the slope was higher for high bog peat than for fen peat.

Equations are often used to describe unsaturated hydraulic conductivity functions. Such expressions provide a method for interpolating or extrapolating hydraulic conductivity curves using limited data, and an efficient data handling procedure for

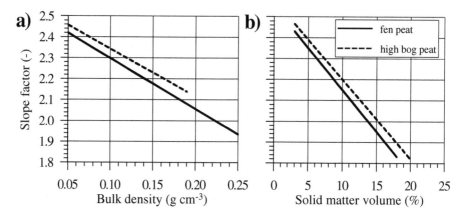

Figure 3.8 Relationship between the slope factor and (a) bulk density and (b) solid matter volume for high bog and fen peats. (Based on data from Bloemen, G.W. 1983. *Zeitschrift für Pflanzenernährung und Bodenkunde*, 146(4):460–473.)

unsaturated flow studies. One of the commonly used equations is the Mualem–Van Genuchten model in the following form:

$$K(h) = K_s \{ [1+ |\alpha h|^n]^m - |\alpha h|^{n-1}|^2 \, /[1+ |\alpha h|^n]^{m(\lambda+2)} \quad (3.4)$$

where **h** is pressure head, and **K** and **K**$_s$ are unsaturated and saturated hydraulic conductivity values, respectively; α, **n** and λ are empirical parameters; and **m** = (1–1/**n**). This equation combines the description of soil water retention characteristic proposed by Van Genuchten (1980) and the pore size distribution model of Mualem (1976). The coefficients in Equation 3.4 are estimated simultaneously from measured soil water retention (α, **n**) and hydraulic conductivity data (**K**$_s$, λ). According to Mualem (1976), the pore connectivity parameter λ averaged about 0.5 across many soils.

Wösten et al. (1999) developed pedotransfer functions to estimate parameters of Equation 3.4 for mineral and organic soils defined according to the FAO soil classification. For organic soils, Wösten et al. (1999) obtained α = 0.0130 cm^{-1}, **n** = 1.2039, λ = 0.40, and **K**$_s$ = 8.0 cm d^{-1}. Calculated unsaturated hydraulic conductivity values using Equation 3.4 and average values for parameters are plotted in Figure 3.9. Measured unsaturated hydraulic conductivity by an evaporation method for fen organic soils as well as functions from Wösten et al. (1999) for mineral soils (coarse and very fine) are also shown. Fen organic soils show high variation in unsaturated hydraulic conductivity: some are close to the curve representing coarse-textured mineral soils, while others are close to the curve representing fine-textured mineral soils. Lundin et al. (1973) examined unsaturated hydraulic conductivity measured by evaporation methods (Korcunov et al., 1961) for several reed-sedge peat samples with similar physical properties (Table 3.4). Data plotted in Figure 3.10 indicate a large range of values of one to two orders of magnitude for peat materials with similar physical properties. A direct measurement of unsaturated hydraulic conduc-

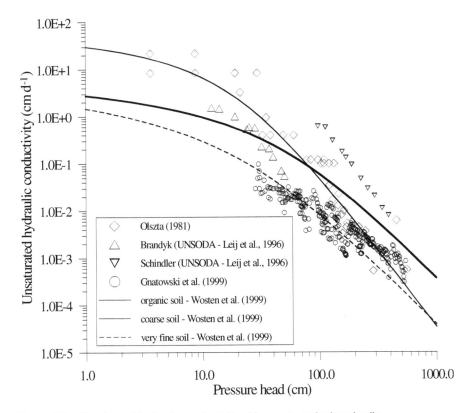

Figure 3.9 Unsaturated hydraulic conductivity of fen peats and mineral soils.

tivity in the field or in the laboratory requires time as well as very precise and expensive equipment. Thus, Bloemen (1983) developed empirical equations relating unsaturated hydraulic conductivity to easily measured peat properties.

Unsaturated hydraulic conductivity can also be derived from parameter identification methods, assuming the reliability of relatively simple experiments. Further assuming algebraic forms of the hydraulic property functions, the water transport

Table 3.4 Physical Properties of Reed-Sedge Peat

Sample	Degree of Decomposition (%)	Sampling Depth (cm)	Bulk density (g cm⁻³)	Ash Content (g kg⁻¹)	Saturated Moisture Content (m³ m⁻³)	Saturated Hydraulic Conductivity (cm d⁻¹)
1	40	50	0.145	75	0.904	103.0
2	40–45	50	0.136	86	0.911	244.0
3	40–45	30	0.158	80	0.897	150.0
4	40–45	40	0.130	78	0.915	350.0
5	40–45	40	0.139	63	0.908	55.0

Source: Based on data from Lundin, K.P., Goncarik, V.M., and Papkievic, I.A. 1973. Examination of water conductivity of unsaturated soil (in Russian), in *Proc. Scientific Research Institute of Land Reclamation and Water Management,* XXI:96–119. Urozaj Press, Minsk, Belarus.

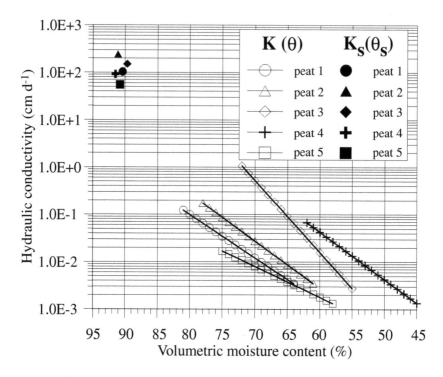

Figure 3.10 Unsaturated hydraulic conductivity functions for reed-sedge peat. (Based on data from Lundin, K.P., Goncarik, V.M., and Papkievic, I.A. 1973. Examination of water conductivity of unsaturated soil (in Russian), in *Proc. Scientific Research Institute of Land Reclamation and Water Management*, XXI:96–119. Urozaj Press, Minsk, Belarus.)

process is simulated and repeated until the simulated results agree with experimental results at the desired degree of accuracy. The inverse parameter estimation method is applied to one-step and multi-step experiments (Van Dam et al., 1994). The one-step and multistep experiments for determining soil moisture retention characteristics and unsaturated hydraulic conductivity functions were investigated in organic soils by Gnatowski et al. (1999). They compared results obtained by one-step and multistep methods with independent measurements using evaporation methods and concluded that the indirect methods yielded satisfactory results in organic soils when the appropriate parameter optimization procedure was used.

VI. SHRINKAGE CHARACTERISTIC

The mechanism and magnitude of volume changes in organic soils due to swelling and shrinkage upon wetting and drying is the result of several forces acting at microscale and leading to soil subsidence and the occurrence of shrinkage cracks (Millette and Broughton, 1984; Szuniewicz, 1989; Gilman, 1994). Subsidence occurs primarily in upper horizons and is most active during the first years following

Table 3.5 Relationship between Shrinkage and Botanical Composition of Fen Peat

Peat Type	No. of Samples	Degree of Decomposition (%)	Bulk Density (g cm^{-3})	Volume of Solid Phase (m^3 m^{-3})	Ash Content (g kg^{-1})	Shrinkage (m^3 m^{-3})
Reed	25	45	0.204	0.109	306	0.65
Sedge-reed	22	40	0.196	0.109	227	0.68
Sedge	29	40	0.156	0.090	176	0.70
Mossy	15	29	0.121	0.071	144	0.59
Sedge-mossy	24	28	0.132	0.077	135	0.60
Alder swamp	27	53	0.212	0.117	260	0.70

Source: Based on data from Ilnicki, P. 1967. *Zeszyty Problemowe Postepow Nauk Rolniczych,* 76:197–311.

drainage. Subsidence can also be caused by loss of buoyancy as a consequence of water removal, and by peat mineralization following soil aeration (Ilnicki, 1973; Gotkiewicz, 1987). Excessive groundwater table drawdown to several meters can generate cracks up to 70 cm wide and 1.3 m deep (Frackowiak and Felinski, 1994). Upon wetting, as cracks close, soil surface rises again. Swelling–shrinkage processes resulting in moorshing and cracking increase water and air conductivity (Okruszko, 1993).

The relationship between water content and volume change in organic soils has been examined in the laboratory (Ilnicki, 1967; Graham and Hicks, 1980; Päivänen, 1982; Szymanowski, 1993b) and in the field (Szuniewicz, 1989; Gilman, 1994; Oleszczuk et al., 1999). Shrinkage as volume change is often expressed as a shrinkage percentage of initial sample volume. Volumetric shrinkage of Polish fen peat materials investigated by Ilnicki (1967) is presented in Table 3.5. The largest shrinkage percentage was observed in sedge and alder peat materials, and the lowest was observed in moss peat. Shrinkage increased with degree of decomposition and decreased with ash content. Shrinkage increased with sampling depth (Päivänen, 1982). The trend for compaction and shrinkage in peat was promoted by the humus component and mitigated by the fiber component (Okruszko, 1960). According to Ilnicki (1967), sapric peat with degree of decomposition over 45% is very susceptible to shrinkage upon drying. Crack formations start at about 0.70 m^3 m^{-3} moisture content, and are well developed at 0.50 m^3 m^{-3}. Woody peat is most susceptible to cracking, followed by sedge peat. Moss peat is the least vulnerable.

Soil moisture and volume relationships have been examined using shrinkage characteristic curves (Stirk, 1954; McGarry and Malafant, 1987), usually by relating void ratio (volume of voids per unit volume of solids) to moisture ratio (volume of water per volume of solids). The curve for heavy clay soils presented in Figure 3.11 shows structural, normal, residual and zero shrinkage phases (Tariq and Durnford, 1993). Structural shrinkage occurs in the wetter range where any soil volume change yields less than water loss by the drainage of large pores, allowing air to enter. Normal shrinkage takes place when the soil volume decrease is equal to water loss as air volume remains constant. In the residual shrinkage phase, water loss exceeds soil volume loss, resulting in more air-filled porosity. In the zero-shrinkage phase, soil volume remains constant, but moisture loss occurs due to pore drainage.

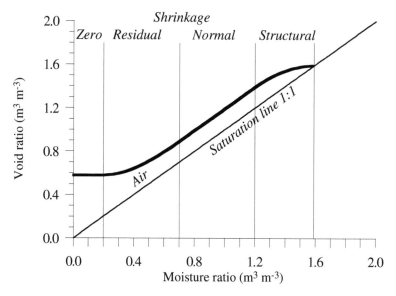

Figure 3.11 Shrinkage curve of a heavy clay soil. (From Tariq, A. and Durnford, D.S. 1993. *Soil Sci. Soc. Am. J.*, 57:1183–1187. With permission.)

Shrinkage data obtained by Szatylowicz et al. (1996) using the saran resin method (Brasher et al., 1966) for hemic peat are shown in Figure 3.12a. The curve fit to three straight lines shows a sudden change in void ratio. Shrinkage is small near saturation, starts with the first water extraction, and increases as soil gets dryer. For segment **c**, the slope is greater than 1 (Figure 3.12a), and volume change is greater than water loss. The apparent wet specific gravity as related to moisture ratio (Figure 3.12b) has one minimum, compared with one maximum for clay soils (Philip, 1969). Similar shapes for shrinkage characteristic curves of organic soils were reported by Van den Akker and Hendriks (1997).

Conversion of soil volume changes into fissuring and subsidence is very important in organic soils because of the large impact of cracks on water and oxygen transport. For converting volume changes into crack volume and subsidence, a dimensionless shrinkage geometry factor (r_s) has been defined as follows (Kim et al., 1992):

$$r_s(0, k) = \log(V_k / V_0) / \log(H_k / H_0) \tag{3.5}$$

where r_s is the shrinkage geometry factor from the initial stage **0** to the **k**-th stage of dryness; V_k and V_0 are soil volumes at stages **k** and **0**, respectively; H_k and H_0 are heights of the soil sample at stages **k** and **0**, respectively. The r_s is equal to 3 for soils with isotropic three-dimensional shrinkage, exceeds 3 when fissuring predominates, is between 1 and 3 for a predominance of subsidence, and is equal to 1 for subsidence only (Bronswijk, 1990).

Examples of shrinkage geometry factors together with corresponding shrinkage characteristic and moisture retention curves are presented in Figure 3.13. Three phases are distinguished from patterns relating the shrinkage geometry factor to

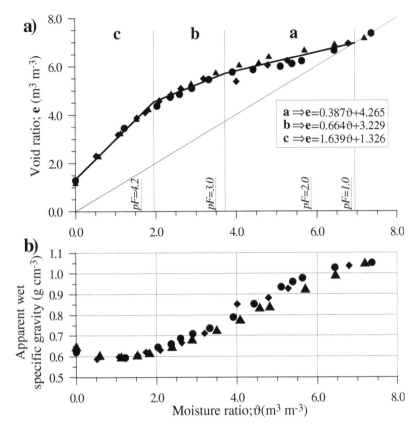

Figure 3.12 Shrinkage curve (a) and the relationship between apparent wet specific gravity and moisture ratio (b) for a mesic tall sedge peat. (From Szatylowicz, J., Oleszczuk, R., and Brandyk, T. 1996. Shrinkage characteristics of some fen peat soils, in *Proc. 10th Int. Peat Congr.*, Bremen, Germany, 2:327–338, With permission.)

moisture ratio. The first phase occurs when soil starts drying from saturation, and the shrinkage geometry factor is equal to 1. The second phase is transitory, and r_s increases as moisture content decreases. In the third phase, r_s is equal to about 3. Cracks may appear when r_s is greater than 1. Bronswijk (1988) applied shrinkage characteristics of heavy clays and the geometry factor to develop equations quantifying the contribution of shrinkage to soil subsidence and estimating crack volume. The model was validated by Oleszczuk et al. (1999) for predicting the thickness and volume change of a peaty moorsh soil under field conditions.

VII. WATER REPELLENCY

A. Methods of Determination

Water repellency has been reported to occur under a wide variety of soil and climatic conditions (Wallis and Horne, 1992). Soil wettability depends on several

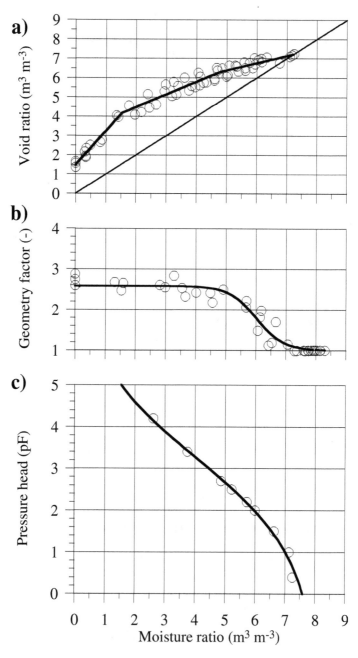

Figure 3.13 Shrinkage curve (a) shrinkage geometry factor (b) and soil moisture retention curve (c) of a sedge peat.

factors mainly related to the quality of organic matter (Ma'shum and Farmer, 1985). In water-repellent soils, water infiltration is reduced, thus accelerating runoff and erosion and impacting plant establishment and growth (DeBano, 1969). Considerable variation in soil water content and irregular moisture patterns were observed in water-repellent mineral soils (Dekker and Ritsema, 1995; Ritsema and Dekker, 1994).

The severity of water repellency is often determined by using two tests: the water drop penetration time (Watson and Letey, 1970) and the alcohol percentage (Dekker and Ritsema, 1994). The water drop penetration time test (WDPT) consists of placing a water drop onto the soil surface, then recording the time for the water drop to penetrate into the sample. The alcohol volume percentage test records the infiltration time of ethanol drops of varying concentrations placed onto the soil surface using a standard medicine dropper. The degree of repellency is the lowest alcohol percentage in the solution penetrating the soil surface within 5 s (Dekker and Ritsema, 1994). These tests can be performed on dried or field-moist samples. Water repellency measured on field-moist samples has been referred to as "actual water repellency," while that measured on dried samples has been called "potential water repellency" (Dekker and Ritsema, 1994).

Water repellency depends on soil moisture content (Berglund and Persson, 1996; De Jonge et al., 1999). The influence of soil moisture content on WDPT of a peat-moorsh soil is presented in Figure 3.14 (Waniek et al., 1999). Dekker and Jungerius (1990) proposed five classes of water repellency: wettable or non-water repellent (WDPT<5 s), slightly (5–60 s), strongly (60–600 s), severely (600–3600 s), extremely water repellent (>3600 s). The WDPT test was sensitive to changes in water content. Moisture content corresponding to pF 2.0, pF 2.7, and pF 4.2 are shown in Figure 3.14. Almost no repellency occurred at soil saturation. Repellency increased rapidly with decreasing water content. Maximum WDPT occurred with dried peat materials in the turf and moorsh layers. The turf layer, or mat, is the top of the moorsh layer densely colonized by plant roots. The WDPT function of soil water content peaked in the alder (40–50 cm) and reed (50–80 cm) peat layers. Highest WDPT occurred near the permanent wilting point (pF 4.2). Under field conditions, where moisture content seldom decreased below the permanent wilting point, the turf layer was classified as nonrepellent, the moorsh layer as slightly repellent, and the alder and reed peat layers as strongly repellent. The persisting potential water repellency of dried samples is presented in Figure 3.15. Generally, peat-moorsh soils present relatively high potential water repellency with up to 0.325 m^3 m^{-3} of ethanol, a level higher than in most mineral soils (Watson and Letey, 1970; McGhie and Posner, 1980).

The liquid–solid contact angle indicates the wettability of a soil (Watson and Letey, 1970). If a drop of water is placed on a hydrophobic or non-wettable surface, it balls up, conferring a large liquid-solid contact angle. In contrast, a water drop placed on a wettable surface spreads over and has a small contact angle. The contact angle for wettable soils is often assumed to be 0°, but for water repellent soils the angle may be larger than 90°. The contact angle is assessed by direct and indirect methods (Wallis and Horne, 1992). Direct methods, the most common, are conducted by placing of a drop of water on the material. The contact angle between a liquid and a solid surface is measured either from droplet profile using an optical goniom-

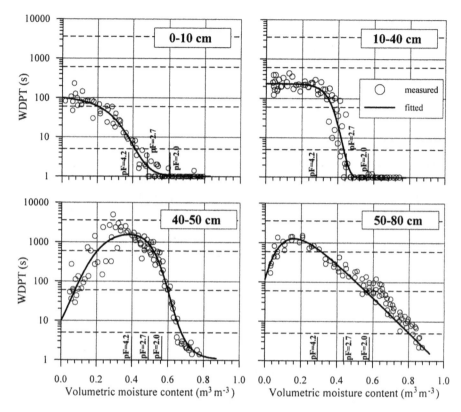

Figure 3.14 Dependence of water repellency on water content at different soil depths across a peat-moorsh soil profile. (From Waniek, E., Szatylowicz, J., and Brandyk, T. 1999. *Roczniki Akademii Rolniczej w Poznaniu CCCX, Melioracje i Inzynieria Srodowiska,* 20(1):199–209. With permission.)

eter, or from droplet geometry (i.e., volume, height, and length). Indirect methods are based on the rate of water movement or the height of capillary rise, which are influenced by liquid–solid contact angles. Equations have been developed for capillary rise (Letey et al., 1962) or rate of water flow through soils (Emerson and Bond, 1963; Hammond and Yuan, 1969), where contact angle is a variable. A contact angle measured by indirect methods is therefore an apparent contact angle.

B. Water Repellency of Organic Soils

Contact angles of organic soil materials were measured by Valat et al. (1991), Lambert and Vanderdeelen (1996), and Holden (1998) using direct methods, and by Waniek et al. (2000) using indirect methods. According to Valat et al. (1991), the solid–liquid contact angle for air-dried peat materials reached 122.1° for woody peat, 116.8° for herbaceous peat and 110.9° for *Sphagnum* peat. Waniek et al. (2000) reported contact angles ranging from 64.2 to 83.1° in peaty moorsh soils developed from fen, using the equilibrium height of capillary rise method; values ranged from

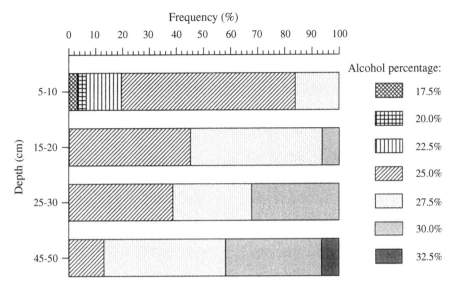

Figure 3.15 Relative frequency of the degree of potential water repellency for dry peat samples at several soil depths. (From Waniek, E., Szatylowicz, J., and Brandyk, T. 1999. *Roczniki Akademii Rolniczej w Poznaniu CCCX, Melioracje i Inzynieria Srodowiska*, 20(1):199–209. With permission.)

86.3 to 89.8° using the dynamic capillary rise approach. Contact angles presented in the literature for organic soils confirm their hydrophobicities or water repellencies.

The influence of water repellency on the variation in soil water content in a field peaty moorsh soil profile was studied by Waniek et al. (1999). Volumetric water content was determined by sampling the soil profile at different depths using steel cylinders (100 cm³ and 5 cm in height). Thirty-one samples were taken from each of the following depths: 5–10, 15–20, 25–30, and 45–50 cm, at close intervals along transects 300 cm in length. Wet soil samples were weighed before oven-drying for 24 h at 105°C, and weighed again to determine water content and bulk density. The soils were sampled on 11 September, 1998 and 19 October, 1998. The spatial distribution of volumetric moisture content is shown in Figure 3.16. The soil moisture profile on September 11 showed a vertically oriented moisture pattern at depths between 10 and 30 cm, and isolated dry spots at 10–30 cm intervals with soil water content less than 40%. In October, the soil was wetter, but the vertical moisture pattern was still present. Presumably, a strongly repellent soil layer at depth of 40–50 cm, as shown in Figure 3.14, restricted capillary rise from groundwater to upper layers. The irregular pattern of spatial distribution of moisture content was typical for soils where preferential water flow occurred (Ritsema et al., 1993).

VIII. SPATIAL VARIABILITY

Gnatowski et al. (1996) examined the spatial variability of bulk density, saturated hydraulic conductivity and moisture content in a peaty moorsh soil covering 5 ha.

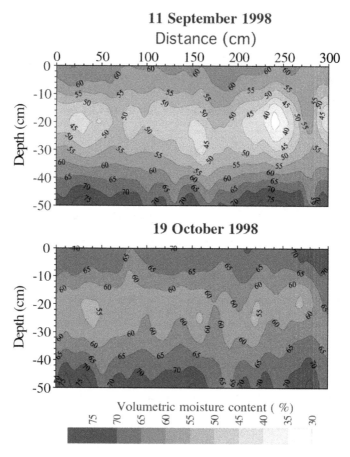

Figure 3.16 Contour plots showing the spatial distribution of volumetric soil water content on September 11 and October 19, 1998 in a Polish peat-moorsh soil. (From Waniek, E., Szatylowicz, J., and Brandyk, T. 1999. *Roczniki Akademii Rolniczej w Poznaniu CCCX, Melioracje i Inzynieria Srodowiska*, 20(1):199–209. With permission.)

Sampling was performed along a 10 m × 10 m grid, totaling 517 nodal points. The results were analyzed using conventional statistics such as mean, maximum, average, median, variance and coefficient of variation (Table 3.6). Coefficients of variation (**CV**) were lowest for saturated and *in situ* moisture contents, intermediate for bulk density and largest for hydraulic conductivity. The ranges of coefficient of variation were similar to those reported for mineral soils (Warrick and Nielsen, 1980). Several soil properties were spatially autocorrelated, thus requiring geostatistical analysis (Webster and Oliver, 1990). The variance of spatially dependent soil parameters was estimated from the semivariogram as follows:

$$\gamma(h) = \frac{1}{2N(h)} \left\{ \sum_{i=1}^{N(h)} [z(x_i + h) - z(x_i)]^2 \right\} \qquad (3.6)$$

where γ is the semivariance, $N(h)$ the number of data pairs separated by lag distance h, and $z(x_1)$, $z(x_2)$, ..., $z(x_n)$ are data for spatial locations x_1, x_2, ..., x_n.

The semivariance computed in Equation 3.6 increases with increasing lag, then levels off. The lag at which a plateau is achieved is the range. The semivariance on the plateau is the sill. Points within the range are considered spatially correlated; points outside the range are declared spatially independent. Empirical semivariograms seldom cross the origin. The semivariance at zero lag is the nugget, composed of two sources of variance: the spatial variance of specimens, at scales less than minimum sampling distance, and local sampling error. The spherical model fits several semivariograms of soil properties and is defined as follows (Webster and Oliver, 1990):

$$\gamma(h) = C_0 + C_s \left[\frac{3}{2} \frac{h}{a} - \frac{1}{2} \left(\frac{h}{a} \right)^3 \right] \qquad 0 < h \le a$$

$$\gamma(h) = C_0 + C_s \qquad\qquad\qquad h > a \qquad\qquad (3.7)$$

$$\gamma(0) = 0$$

where C_0 is the nugget effect, C_s is the structural component, a is the range of spatial dependence, and $(C_0 + C_s)$ is total variance (the sill). The linear model without sill is defined as follows:

$$\gamma(h) = C_0 + wh \qquad 0 < h$$

$$\gamma(0) = 0 \qquad\qquad\qquad\qquad\qquad (3.8)$$

where w is the slope and all other symbols are as defined previously.

Gnatowski et al. (1996) applied spatial statistics to assess correlations among peat properties (bulk density, moisture content, and saturated hydraulic conductivity) across a field of 5 ha (Figure 3.17) in a peaty moorsh soil. For saturated and *in situ*

Table 3.6 Statistical Analysis of Some Physical Peaty Moorsh Properties

Property	Value					CV[a]
	Minimum	Maximum	Average	Median	Variance	(%)
Bulk density (g cm^{-3})	0.206	0.559	0.274	0.263	0.0021	16.1
Saturated volumetric moisture content (m^3 m^{-3})	0.738	0.938	0.841	0.841	8.27	3.4
In situ volumetric moisture content (m^3 m^{-3})	0.538	0.852	0.748	0.753	19.59	5.9
Saturated hydraulic conductivity (m d^{-1})	0.01	2.59	0.26	0.13	0.142	142.4

[a] Coefficient of variation.

Source: From Gnatowski, T., Brandyk, T., and Szatylowicz, J. 1996. *Przeglad Naukowy Wydzialu Melioracji i Inzynierii Srodowiska*, 11:129–136. With permission.)

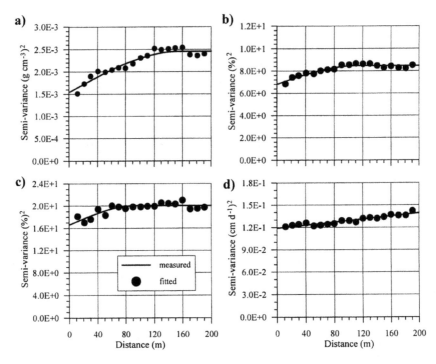

Figure 3.17 Experimental and fitted semivariograms for (a) bulk density, (b) saturated moisture content, (c) *in situ* moisture content, and (d) saturated hydraulic conductivity. (From Gnatowski, T., Brandyk, T., and Szatylowicz, J. 1996. *Przeglad Naukowy Wydzialu Melioracji i Inzynierii Srodowiska*, 11:129–136. With permission.)

moisture contents and bulk density values, empirical semivariograms were fitted using spherical models while, for saturated hydraulic conductivity, a linear model was used (Table 3.7). The smallest range was observed for *in situ* moisture content (94 m), and the highest for bulk density (157 m). Optimized semivariogram parameters (Table 3.7) showed that the nugget component (C_0) for saturated and *in situ*

Table 3.7   Parameters Derived from the Estimated Semivariogram Models for Physical Peaty Moorsh Properties in the Biebrza Valley, Poland

Property	Model	Parameters of Semivariogram			$C_0 (C_0 + C_s)$ (%)
		C_0	C_s	a (m)	
Bulk density (g cm⁻³)	Spherical	0.00154	0.000918	157.0	62.6
Saturated volumetric moisture content (%)	Spherical	6.81	1.66	112.0	80.4
Actual volumetric moisture content (%)	Spherical	16.6	3.43	94.1	82.9
Saturated hydraulic conductivity (m d⁻¹)	Linear	0.119	0.000102[a]		

[a] Parameter **w** in linear model Equation 3.8.

Source: Based on data from Gnatowski, T., Brandyk, T., and Szatylowicz, J. 1996. *Przeglad Naukowy Wydzialu Melioracji i Inzynierii Srodowiska*, 11:129–136.

Table 3.8   Statistical Analysis of Moorsh Retention Data in the Biebrza Valley, Poland

Measures	Volumetric Moisture Content (m³ m⁻³) at Pressure Head (pF)						
	0.4	1.0	1.5	1.8	2.0	2.7	4.2
Mean	0.8094	0.7970	0.7634	0.7287	0.6913	0.5925	0.5709
Median	0.8103	0.8007	0.7668	0.7323	0.6967	0.5934	0.5626
Maximum	0.8556	0.8376	0.8221	0.7900	0.7561	0.7074	0.6944
Minimum	0.7326	0.7231	0.6809	0.6438	0.6065	0.5078	0.4792
Variance ($\times 10^{-4}$)	7.13	6.92	10.76	13.25	13.62	19.98	25.60
CV[a] (%)	3.30	3.30	4.30	5.00	5.34	7.55	8.86

[a] Coefficient of variation.

moisture contents and for bulk density varied from 63 to 83% of the sill $(C_0 + C_s)$, thus indicating a significant contribution of local variation in peat properties between sampling points (10 m).

Spatial variability of moisture retention characteristics was examined along a 50-m transect located in the Middle Biebrza Basin in Poland. The soil was classified as a hemic peaty moorsh organic soil. The stand was a permanent grassland. Soil samples were collected 1 m apart in the moorsh layer (0.10 – 0.20 m). Soil moisture retention was measured using a standard sand table (pF 0.0–2.0) and pressure chambers (pF 2.7–4.2) (Klute, 1986). The Van Genuchten (1980) model was fitted to the 51 pF curves using the RETC parameter optimization code (Van Genuchten et al., 1991).

Statistical parameters for water content at various matric potentials are presented in Table 3.8. Generally, the **CV**s did not exceed 10% and increased as matric potential became more negative (i.e., drier conditions). The mean, median, variance, and **CV** are presented in Table 3.9 for each Van Genuchten parameter. The **CV**s were 3.2% for θ_s, 9.4% for θ_r, and 10.5% for **n**. The parameter α describing the shape of the moisture retention curve showed the highest **CV** (75.8%). The range and trend of **CV** values for the organic soil were similar to those reported for mineral soils (Mallants et al., 1995).

For each parameter of the Van Genuchten equation, experimental semivariograms were computed. Spherical semivariograms are presented in Figure 3.18. Semivario-

Table 3.9   Statistical Analysis of the Parameters in the Van Genuchten model for Describing Water Retention Characteristic in a Moorsh Soil of the Biebrza Valley, Poland

Measures	Parameters			
	θ_s (m³ m⁻³)	θ_r (m³ m⁻³)	α (cm⁻¹)	n (–)
Mean	0.8224	0.5615	0.0322	1.7839
Median	0.8195	0.5486	0.0236	1.7499
Maximum	0.8671	0.6925	0.1281	2.2772
Minimum	0.7513	0.4714	0.0108	1.4630
Standard deviation	0.0262	0.0529	0.0244	0.1882
CV[a] (%)	3.18	9.41	75.77	10.55

[a] Coefficient of variation.

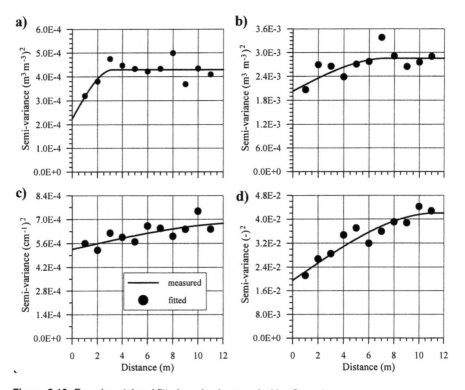

Figure 3.18 Experimental and fitted semivariograms for Van Genuchten parameters describing water retention characteristics for moorsh soil: (a) θ_s, (b) θ_r, (c) α and (d) n.

gram parameters are summarized in Table 3.10. The smallest range was observed for saturated moisture content (3.21 m); for α, it was the highest (14.2 m). The nugget component (C_0) accounted for 46.4% of the sill (lowest value) for n and 77.1% (highest value) for α. The nugget effect indicated significant variation between sampling points (1 m). Using the kriging method (Webster and Oliver, 1990), contour maps were obtained for saturated and *in situ* moisture content, bulk density, and saturated hydraulic conductivity for a peat-moorsh organic soil over an area of 5 ha (Gnatowski et al., 1996) as illustrated in Figure 3.19.

Table 3.10 Parameter of the Estimated Spherical Semi-variogram for Van Genuchten Parameters Describing Soil Water Retention Characteristics in Moorsh Soils of the Biebrza Valley, Poland

Parameter	Parameters of Semivariogram			$C_0 (C_0 + C_s)$ (%)
	C_0	C_s	a (m)	
θ_s (m^3 m^{-3})	2.25 10^{-4}	2.04 10^{-4}	3.21	52.4
θ_r (m^3 m^{-3})	20.30 10^{-4}	8.22 10^{-4}	7.43	71.2
α (cm^{-1})	5.26 10^{-4}	1.56 10^{-4}	14.20	77.1
n	195.00 10^{-4}	225.00 10^{-4}	11.40	46.4

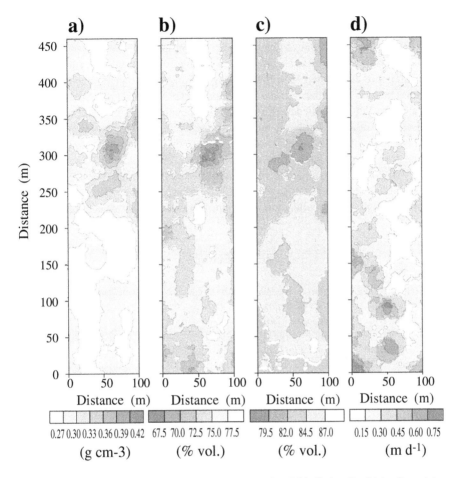

Figure 3.19 Contour maps of peat-moorsh soil properties: (a) bulk density, (b) *in situ* moisture contents, (c) saturated moisture content and (d) saturated hydraulic conductivity. (From Gnatowski, T., Brandyk, T., and Szatylowicz, J. 1996. *Przeglad Naukowy Wydzialu Melioracji i Inzynierii Srodowiska,* 11:129–136. With permission.)

IX. CONCLUSION

In organic soils, water retention and hydraulic conductivity depend on the degree of peat decomposition and botanical origin. Peat decomposition markedly influences plant-available water and soil drainable porosity. Saturated hydraulic conductivity is highly variable and influenced by soil swelling–shrinkage cycles as well as the sealing of soil pores by gas bubbles and other products of organic matter decay produced by anaerobic microbiological processes. In some peat deposits, hydraulic conductivity is anisotropic. A limited number of measured, unsaturated hydraulic conductivity data are reported in the literature, mainly due to their laborious determination procedures. Generally, unsaturated hydraulic conductivity is more variable than saturated hydraulic conductivity. For organic soils, only a few attempts have been made to develop empirical functions (so-called pedotransfer functions) to estimate parameters of the models describing soil water retention and unsaturated hydraulic conductivity from easily measured soil properties.

Most organic soils tend to shrink and become hydrophobic upon drying. Changes in soil moisture are related to volume changes using shrinkage characteristics and geometry factors. The shape of the shrinkage curve is different in organic than in mineral soils. Water repellency of organic soils depends on soil water content and increases variability in field soil moisture patterns. Any water repellent layer can have a negative influence on soil moisture content by reducing the amount of water supplied through infiltration from the soil surface and from the groundwater by capillary rise. In order to simulate water transport and oxidation processes in organic soils, shrinkage characteristics together with shrinkage geometry factors and parameters describing water repellency must be incorporated into hydrological models.

REFERENCES

Amoozegar, A. and Warrick, A.W. 1986. Hydraulic conductivity of saturated soils: Field methods, in *Methods of Soil Analysis, Part 1,* 2nd ed. Agron. Monogr. 9, Soil Sci. Soc. Am., Klute, A., Ed., Madison, WI. 735–770.

Baden, W. and Eggelsmann, R. 1963. Permeability of organic soils (in German). *Zeitschrift für Kulturtechnik und Flurbereinigung,* 4:226–254.

Bartels, R. and Kuntze, H. 1973. Peat properties and unsaturated hydraulic conductivity of organic soils (in German). *Zeitschrift für Pflanzenernährung und Bodenkunde,* 134(2):125–135.

Berglund K. and Persson, L. 1996. Water repellency of cultivated organic soils. *Acta. Agric. Scand.* 46:145–152.

Bloemen, G.W. 1983. Calculation of hydraulic conductivities and steady state capillary rise in peat soils from bulk density and solid matter volume. *Zeitschrift für Pflanzenernährung und Bodenkunde,* 146(4):460–473.

Boelter, D.H. 1965. Hydraulic conductivity of peats. *Soil Sci.,* 100:227–231.

Boelter, D.H. 1969. Physical properties of peats as related to degree of decomposition. *Soil Sci. Soc. Am. Proc.,* 33:606–609.

Bondarenko, N.F., Danchenko, O.I., and Kovalenko, N.P. 1975. Studying the nature of filtration flow instability in peat (in Russian with English summary). *Pochvovedenie (Soviet Soil Science),* 10:137–139.

Brandyk, T., Gnatowski, T., and Szatylowicz, J. 1996. Spatial variability of some physical properties of decomposed lowland peat soil. *Proc. 10th Int. Peat Congr.,* 2:294–305.

Brasher, B.R. et al. 1966. Use of Saran resin to coat natural soil clods for bulk density and water-retention measurements. *Soil Sci.,* 101:108.

Bronswijk, J.J.B. 1988. Modeling of water balance, cracking and subsidence of clay soils. *J. Hydrol.,* 97:199–212.

Bronswijk, J.J.B. 1990. Shrinkage geometry of a heavy clay soil at various stresses. *Soil Sci. Soc. Am. J.,* 54:1500–1502.

Chason, D.B. and Siegel, D.I. 1986. Hydraulic conductivity and related physical properties of peat, Lost River peatlands, northern Minnesota. *Soil Sci.,* 142:91–99.

De Jonge, L.W., Jacobsen, O.H. and Moldrup, P. 1999. Soil water repellence: Effects of water content, temperature, and particle size. *Soil Sci. Soc. Am. J.,* 63:437–442.

DeBano, L.F. 1969. Water repellent soils: a worldwide concern in management of soil and vegetation. *Agric. Sci. Rev.,* 7(2):11–18.

Dekker, L.W. and Ritsema, C.J. 1994. How water moves in a water repellent sandy soil. 1. Potential and actual water repellence. *Water Resour. Res.,* 30:2507–2517.

Dekker, L.W. and Ritsema, C.J. 1995. Fingerlike wetting patterns in two water-repellent loam soils. *J. Environ. Qual.,* 24:324–333.

Dekker, L.W. and Jungerius, P.D. 1990. Water repellence in the dunes with special reference to The Netherlands. *Catena,* Suppl. 18:173–183.

Dirksen, C. 1991. Unsaturated hydraulic conductivity, in *Soil Analysis. Physical Methods.* Smith, A.S. and Mullins, C.E., Eds., Marcel Dekker, New York, 209–270.

Emerson, W.W. and Bond, R.D. 1963. The rate of water entry into dry sand and calculation of the advancing contact angle. *Austr. J. Soil Res.,* 1:9–16.

Frackowiak, H. and Felinski, T. 1994. Subsidence of the surface of meadow organic soils under intensive drying conditions (in Polish with English summary). *Wiadomosci Instytutu Melioracji i Uzytkow Zielonych,* XVIII(2):29–36.

Gilman, K. 1994. *Hydrology and Wetland Conservation.* John Wiley & Sons, Chichester, UK.

Gnatowski, T., Brandyk, T., and Szatylowicz, J. 1996. Spatial variability estimation of some soil properties at irrigated plot scale (in Polish with English summary). *Przeglad Naukowy Wydzialu Melioracji i Inzynierii Srodowiska,* 11:129–136.

Gnatowski, T., Brandyk, T., Szatylowicz, J., and Oleszozuk, R. 1999. Parameter estimation by one-step and multi-step outflow experiments for peat soils, in *Modeling of Transport Processes in Soils at Various Scales in Time and Space.* Feyen, J. and Wiyo, K., Eds., Proc. Int. Work. EurAgEng's, Field of Interest on Soil and Water, Leuven, Belgium, 206–214.

Gotkiewicz, J. 1987. Mineralization of organic nitrogen in peat-muck soils in long-term experiments (in Polish). *Bibl. Wiadomosci Instytutu Melioracji i Uzytkow Zielonych,* 68:85–98.

Graham, R.B. and Hicks, W.D. 1980. Some observations on the changes in volume and weight of peat and peat moss with loss of moisture, in *Proc. 6th Int. Peat Congr.,* Duluth, MN, 554–558.

Hammond, L.C. and Yuan, T.L. 1969. Methods of measuring water repellence of soils, in *Proc. Symp. Water Repellent Soils.* DeBano, L.F. and Letey, J., Eds., Univ. California, Riverside, CA, 49–59.

Hemond, H.F. and Goldman, J.C. 1985. On non-Darcian water flow in peat. *J. Ecol.,* 73:579–584.

Holden, N.M. 1998. By-pass of water trough laboratory columns of milled peat. *Int. Peat J.,* 8:13–22.

Illner, K. and Raasch, H. 1977. Influence of peat properties on unsaturated hydraulic conductivity of fen soils (in German). *Archiv für Acker Pflanzenbau und Bodenkunde,* 21(10): 753–758.

Ilnicki, P. 1967. Shrinkage of peat soils during drying depends on soil structure and physical properties (in Polish). *Zeszyty Problemowe Postepow Nauk Rolniczych,* 76:197–311.

Ilnicki, P. 1973. Subsidence rate of reclaimed peatlands in the Notec River Valley (in Polish). *Zeszyty Problemowe Postepow Nauk Rolniczych,* 146:33–61.

Ingram, H.A.P., Rycroft, D.W., and Williams, D.J.A. 1974. Anomalous transmission of water through certain peats. *J. Hydrol.,* 22:213–218.

Ivanov, K.E., 1953. *Wetland Hydrology* (in Russian). Gidrometeoizdat Press, Leningrad (Sankt Petersburg), Russia.

Kim, D.J., Vereecken, H., Feyen, J., Bods, D., and Bronswyk, J.J.B. 1992. On the characterization of properties of an unripe marine clay soil, I. Shrinkage processes of an unripe marine clay soil in relation to physical ripening. *Soil Sci.,* 153:471–481.

Klute, A. 1986. Water retention: Laboratory methods, in *Methods of Soil Analysis. Part 1,* 2nd ed. Agron. Monogr. 9, Klute, A., Ed., Soil Sci. Soc. Am., Madison, WI, 635–662.

Klute, A. and Dirksen, C. 1986. Hydraulic conductivity and diffusivity: Laboratory methods, in *Methods of Soil Analysis, Part 1,* 2nd ed. Agron. Monogr. 9, Klute, A., Ed., Soil Sci. Soc. Am., Madison, WI, 687–734.

Korcunov, S.S., Mogilevski, I.I., and Abakumov, O.N. 1961. Determination of unsaturated conductivity by steady-state evaporation from the soil surface (in Russian), in *Proc.Central Scientific Research Institute of Peat Industry,* XVII: 156–166, National Energy Press, Leningrad (Sankt Peterburg), Russia.

Korpijaakko, M. 1988. Consideration of the factors affecting the hydraulic conductivity of peat as ground of both laboratory and field tests, in *Proc. 8th Int. Peat Congr.,* 3: 127–136, Leningrad (Sankt Peterburg), Russia.

Korpijaakko, M. and Radforth, N.W. 1972. Studies on the hydraulic conductivity of peat. *Proc. 4th Int. Peat Congr.,* 3:323–334.

Kramer, W. and Meyer, B. 1968. Measurement of the unsaturated hydraulic conductivity of soil cores in their natural state using the double-membrane apparatus (in German). *Göttinger Bodenkundliche Berichte,* 1:127–154.

Lambert, K. and Vanderdeelen, J. 1996. Reliable analytical techniques for the evaluation of physical and chemical properties of lowland tropical peat soils. *Int. Peat J.,* 6:50–76.

Leij, F.J. et al. 1996. Unsaturated soil hydraulic database, UNSODA 1.0 User's Manual. USEPA Rep. EPA/600/R-96/095, Ada, Oklahoma.

Letey, J., Osborn, J., and Pelishek, R.E. 1962. Measurement of liquid-solid contact angles in soil and sand. *Soil Sci.,* 93:149–153.

Lishtvan, I.I., Bazin, E.T., and Kosov, V.I. 1989. *Physical Processes in Peat Deposits* (in Russian). Nauka i Tehnika Press, Minsk, Belarus.

Lundin, K.P. 1964. *Water Properties of Peat Deposits* (in Russian). Urozaj Press, Minsk, Belarus.

Lundin, K.P., Goncarik, V.M., and Papkievic, I.A. 1973. Examination of water conductivity of unsaturated soil (in Russian), in *Proc. Scientific Research Institute of Land Reclamation and Water Management,* XXI: 96–119. Urozaj Press, Minsk, Belarus.

Ma'shum, M. and Farmer, V.C. 1985. Origin and assessment of water repellence of a sandy South Australian soil. *Austr. J. Soil Res.,* 23:623–626.

Mallants, D., Mohanty, B.P., Jacques, D., and Feyen, J. 1995. Spatial variability of hydraulic properties in a multi-layered soil profile. *Soil Sci.,* 161:167–181.

McGarry, D. and Malafant, K.W.J. 1987. The analysis of volume change in unconfined units of soil. *Soil Sci. Soc. Am. J.,* 51 290–297.

McGhie, D.A. and Posner, A.M. 1980. Water repellence of a heavy–textured Western Australian surface soil. *Austr. J. Soil. Res.,* 18:309–323.

Messing, I. and Jarvis. 1990, N.J. Seasonal variation in field-saturated hydraulic conductivity in two swelling clay soils in Sweden. *J. Soil. Sci.,* 41:229–237.

Millette, J.A. and Broughton, R.S. 1984. The effect of water table in organic soil on subsidence and swelling. *Can. J. Soil Sci.,* 64:273–282.

Mualem, Y. 1976. A new model for predicting the hydraulic conductivity of unsaturated porous media. *Water Resour. Res.,* 12:513–522.

Okruszko, H. 1960. Muck soils of valley peatlands, and their chemical and physical properties. *Roczniki Nauk Rolniczych Ser.,* F74:5–89.

Okruszko, H. 1971. Determination of specific gravity of hydrogenic soils on the basis of their mineral particles content (in Polish with English summary). *Wiadomosci Instytutu Melioracji i Uzytkow Zielonych,* X(1):47–54.

Okruszko, H. 1976. Keys to hydrogenic soil investigation and classification for reclamation purposes (in Polish). *Bibl. Wiadomosci Instytutu Melioracji i Uzytkow Zielonych,* 52: 7–54.

Okruszko, H. 1993. Transformation of fen-peat soils under the impact of draining. *Zeszyty Problemowe Postepow Nauk Rolniczych*, 406:3–73.

Oleszczuk, R., Szatylowicz, J., and Brandyk, T. 1995. Estimation of the hydraulic properties in peaty moorsh soil with medium degree of moorshing (in Polish with English summary). *Przeglad Naukowy Wydzialu Melioracji i Inzynierii Srodowiska*, 7:11–20.

Oleszczuk, R., Szatylowicz, J., and Brandyk, T. 1999. Prediction of vertical peaty moorsh soil profile movements caused by moisture content changes. *Roczniki Akademii Rolniczej w Poznaniu CCCX, Melioracje i Inzynieria Srodowiska*, 20(1):139–150.

Olszta, W. 1981. Laboratory investigations of the capillary conductivity of soils (in Polish with English summary). *Wiadomosci Instytutu Melioracji i Uzytkow Zielonych*, XIV(2):201–213.

Ostromecki, J. 1936. About some functional relationships between physical properties of peat and peatlands (in Polish with German summary). Ph.D. Dissertation, published in *Roczniklakowy i Torfowy*.

Päivänen, J. 1973. Hydraulic conductivity and water retention in peat soils. *Acta For. Fenn.*, 129:1–70.

Päivänen, J. 1982. Physical properties of peat samples in relation to shrinkage upon drying. *Silva Fennica*, 16(3):247–263.

Philip, J.R. 1969. Moisture equilibrium in the vertical in swelling soils, I. Basic theory. *Austr. J. Soil. Res.*, 7:99–120.

Renger, M. et al. 1976. Kapillarer Aufsteig aus dem Grundwasser und Infiltration bei Moorböden (in German). *Geologische Jahrbuch*, F3:9–51.

Richards, L.A. and Wilson, B.D. 1936. Capillary conductivity measurements in peat soils. *J. Am. Soc. Agron.*, 28:427–431.

Rijtema, P.E. 1969. *Soil Moisture Forecasting. Nota 513, ICW*, Wageningen, The Netherlands.

Ritsema, C.J. and Dekker, L.W. 1994. Soil moisture and dry bulk density patterns in bare dune sands. *J. Hydrol.*, 154:107–131.

Ritsema, C.J., Dekker, L.W., Hedrickx, J.M.H., and Hamminga, W. 1993. Preferential flow mechanism in a water repellent sandy soil. *Water Resour. Res.*, 29:2183–2193.

Rycroft, D.W., Williams, D.J.A., and Ingram, H.A.P. 1975a. The transmission of water through peat, I. Review. *J. Ecol.*, 63:535–556.

Rycroft, D.W., Williams, D.J.A., and Ingram, H.A.P. 1975b. The transmission of water through peat, II. Field experiments. *J. Ecol.*, 63:557–568.

Stirk, G.B. 1954. Some aspects of soil shrinkage and the effect of cracking upon water entry into the soil. *Aust. J. Agric. Res.*, 5:279–290.

Szatylowicz, J., Oleszczuk, R., and Brandyk, T. 1996. Shrinkage characteristics of some fen peat soils, in *Proc. 10th Int. Peat Congr.*, Bremen, Germany, 2:327–338.

Szuniewicz, J. 1989. Pulsation retention and formation of water conditions in dry years in reclamated sedge-moss peats (in Polish with English summary). *Wiadomosci Instytutu Melioracji i Uzytkow Zielonych*, XVI(2):169–183, Falenty, Poland.

Szymanowski, M. 1993a. Basic physico-hydrological and retention properties and their relationship with bulk density of various weakly-sludged (low-ash) peat formations (in Polish with English summary). *Wiadomosci Instytutu Melioracji i Uzytkow Zielonych*, XVII(3):153–174, Falenty, Poland.

Szymanowski, M. 1993b. Shrinkage of low-ash peat soils and evaluation of moisture under the field conditions, using curves of water sorption (in Polish with English summary). *Wiadomosci Instytutu Melioracji i Uzytkow Zielonych*, XVII(3):175–189, Falenty, Poland.

Tariq, A. and Durnford, D.S. 1993. Analytical volume change model for swelling clay soils. *Soil Sci. Soc. Am. J.*, 57:1183–1187.

Tietje, O. and Tapkenhinrichs, M. 1993. Evaluation of pedo-transfer functions. *Soil Sci. Soc. Am. J.*, 57:1088–1095.

Valat, B., Jouany, C., and Riviere, L.M. 1991. Characterization of the wetting properties of air-dried peats and composts. *Soil Sci.*, 152:100–107.

Van Dam, J.C., Stricker, J.N.M., and Droogers, P. 1994. Inverse method to determine soil hydraulic functions from multistep outflow experiments. *Soil Sci. Soc. Am. J.*, 58:647–652.

Van den Akker, J.J.H. and Hendriks, R.F.A. 1997. Shrinkage characteristics of Dutch peat soils. *Proc. Int. Conf. Peat, Horticulture — Its Use and Sustainability*, 156–163, Amsterdam, The Netherlands.

Van Genuchten, M.Th. 1980. A closed-form equation for predicting the hydraulic conductivity of unsaturated soils. *Soil Sci. Soc. Am. J.*, 44:892–898.

Van Genuchten, M.Th., Leij, F.J., and Yates, S.R. 1991. The RETC code for quantifying the hydraulic functions of unsaturated soils. USEPA Rep. EPA/600/2–91/065, Ada, Oklahoma.

Von Post, L. 1922. Swedish geological peat survey with the results obtained so far (in Swedish). *Svenska Mosskulturföreningens tidskrift*, 36:1–27.

Wallis, M.G. and Horne, D.J. 1992. Soil water repellence. *Adv. Soil Sci.*, 20 91–146.

Waniek, E., Szatylowicz, J., and Brandyk, T. 1999. Moisture patterns in water repellent peat-moorsh soil. *Roczniki Akademii Rolniczej w Poznaniu CCCX, Melioracje i Inzynieria Srodowiska*, 20(1):199–209.

Waniek, E., Szatylowicz, J., and Brandyk, T. 2000. Determination of soil-water contact angles in peat-moorsh soils by capillary rise experiments. *Suo*, 51:149–154.

Warrick, A.W. and Nielsen, D.R. 1980. Spatial variability of soil physical properties in the field, in *Applications of Soil Physics*. Hillel, D., Ed., Academic Press, New York, 319–344.

Watson, C.J. and Letey, J. 1970. Indices for characterizing soil-water repellence based upon contact angle-surface tension relationships. *Soil Sci. Soc. Am. Proc.*, 34: 841–844.

Webster, R. and Oliver, M.A. 1990. *Statistical Methods in Soil and Land Resource Survey*. Oxford University Press, U.K.

Weiss, R., Alm, J., Laiho, R., and Laine, J. 1998. Modeling moisture retention in peat soils. *Soil Sci. Soc. Am. J.*, 62:305–313.

Wind, G.P. 1969. Capillary conductivity data estimated by a simple method. *Proc. UNESCO/International Association of Hydrological Sciences Symp., Water in the Unsaturated Zone*, 181–191, Wageningen, The Netherlands.

Wösten, J.H.M., Lilly, A., Nemes, A., and LeBas, C. 1999. Development and use of a database of hydraulic properties of European soils. *Geoderma*, 90:169–185.

CHAPTER 4

Quality of Peat Substrates for Plants Grown in Containers

Jean Caron and Louis-Marie Rivière

CONTENTS

ABSTRACT

Peat materials are widely used in growing media for nursery and greenhouse productions. Tillage of peat fields as well as harvesting, piling, screening, crushing, and handling operations influence particle size distribution and peat quality without influencing decomposition degree. Quality attributes of peat-base products are physical (air–water storage and exchange), and chemical (pH, salinity, and nutrient content). Determination methods are briefly described. Total porosity, air-filled porosity, storage capacity for available water, and container capacity are associated with peat degree of decomposition and botanical makeup. Pore tortuosity and gas diffusivity are more closely related to grain-size distribution and plant productivity than capacity criteria such as air-filled porosity and volume of available water. Structuring components to be combined with peat materials should increase gas diffusion while maintaining high amounts of available water. Nutrient availability is primarily related to substrate pH, salinity, and cation exchange capacity. Fibric and hemic *Sphagnum* peat and some herbaceous peat materials are used successfully in peat mixes.

I. INTRODUCTION

Nursery and greenhouse crops are grown in artificial mixes to avoid soilborne plant pathogens and maintain high transpiration rates. Peat, alone or in combination with other materials, is the material most widely used as growing medium, despite substantial research aimed at replacing it with industrial wastes (Schmilewski, 1992). Peat production is evaluated at about 25,000,000 m^3 yr^{-1} worldwide, mostly in Canada and Europe (95%). The main buyers are the United States (5,800,000 m^3 yr^{-1}) and The Netherlands (2,500,000 m^3 y^{-1}) (Hood, 1998).

Physiological, chemical, and physical processes drive the flux of solutes and water at the soil–root interface. The substrate quantitatively regulates the amount and transfer of water in relation to water potential gradients. Substrate management is qualitative when based on solute or gas concentrations. Solute concentration at any time or location is the result of complex interactions between the solid and liquid phases involving chemical (precipitation, dissolution, ion exchange at the solid surfaces, convective, and diffusive transport) and biochemical (mineralization, reorganization) processes. Gas concentrations depend on the exchange of oxygen and carbon dioxide between the liquid and gas phases, and within the gas phase.

In order to produce high-quality, substrate-sustaining plant production in containers, substrate criteria are defined for important physical and chemical properties, and their methods of determination are described in this chapter.

II. PEAT SUBSTRATE QUALITY: BASIC PROPERTIES

A. Bulk Density

Artificial media, especially peat, are sensitive to settling and loosening caused by substrate handling (Heiskanen et al., 1996). Large variations in the bulk density of primary components excludes the mass basis to make substrates, unless bulk density is indicated. A reliable approach to measure bulk density is of great importance because standard methods are based on the relationship between bulk density and plant growth (Bragg, 1997). Sample preparation should yield bulk density values close to cultivated substrates. The method must be repeatable, accurate, and inexpensive.

In Europe, standardized methods use large-volume samples (20 L) prepared without compaction (e.g., CEN method 12580), yielding relatively low bulk density values for expanding materials such as white *Sphagnum* peat (Morel et al., 1999). A 30-cm diameter 20-L cylinder is equipped with an uplifting cylinder 7.5 cm high with the same diameter at the top of which a 2-, 4- or a 6-mm sieve is attached, depending on coarseness of the material. The cylinder is fed through the sieve to loosen the material and obtain a constant fall height. Substrate is levelled on top of the cylinder after removing the sieve and uplifting the cylinder. In North America, such standardized methods have not been adopted yet, and smaller volumes (usually less than 1 L) are used. The substrate is naturally drained after saturation, and an external pressure is applied onto it in addition to the overburden pressure (Hidding, 1999). With the EN-13041 method, the sample is saturated, equilibrated at −5 kPa on a tension table, transferred into double ring cylinders, rewetted, and equilibrated at −1 kPa (Gabriels, 1995). Samples can be equilibrated on tension tables or in Tempe cells (Parent and Caron, 1993; Caron et al., 1997). These methods are intended to mimic natural settling in potted substrates under cultivation.

Techniques have been developed to measure bulk density directly in potted substrates without any substrate handling (Paquet et al., 1993). After compaction, either natural or artificial, bulk density is determined by using gravimetric or time domain reflectometry (TDR) techniques (Topp et al., 1980). With TDR, bulk density is deduced from total porosity (water content after complete rewetting of the substrate) and particle density. The TDR apparatus determines the apparent dielectric constant of the medium, from which volumetric water content is derived. The apparatus must be calibrated *a priori* against volumetric water content determined by weighing, because equations derived for mineral soils do not apply to organic soils (Paquet et al., 1993). Paquet et al. (1993), Anisko et al. (1994) and Da Sylva et al. (1998) published calibration equations for various organic-mineral soil mixtures. With this technique for water retention measurements, rewetting from underneath by gradual elevation of the water level and immersion for 24 h is successfully used for moist fibric peat (Caron et al., 2001), but may lead to serious errors with air-dried fibric or sapric peats (Wever, 1995).

Table 4.1   Semi-Empirical Regression Equations to Calculate Total Porosity (TP in % v/v) from Dry Bulk Density (ρ_{db} in g cm⁻³)

Equation	R²	Reference
TP = 95.83 – 32.43 ρ_{db}	0.97	Gras (1982)
TP = 94.1 – 32.8 ρ_{db}	0.92	Beardsell et al. (1979)
TP = 96.72 – 38.1 ρ_{db}	0.97	Waller and Harrison (1991)
TP = 98.39 – 36.55 ρ_{db}	0.99	Bunt (1984)

B. Porosity

Total porosity (*TP*) of a substrate is defined as the ratio of voids (V_v) occupied by fluids (liquids and gases) to total volume V_t, which is the volume of voids V_v plus the volume of the solid phase V_s. Total porosity can be computed from dry bulk density (ρ_{db}) and particle density (ρ_p) as follows:

$$TP = 1 - \frac{\rho_{db}}{\rho_p} \tag{4.1}$$

where ρ_{db} is oven-dry mass (105°C) divided by initial volume.

Particle density is determined using a water or kerosene pycnometer, or from an assumed particle density (Waller and Harrison, 1991) computed as follows:

$$\rho_p = \frac{1}{\dfrac{F}{2.65} + \dfrac{1-F}{1.5}} \tag{4.2}$$

where *F* is the ash content in g per g of dry substrate (Lemaire et al., 1989). Semi-empirical formulae also exist for total porosity (Table 4.1). Alternatively, total porosity can be determined directly by inserting TDR probes into a saturated substrate directly in the pot (Paquet et al., 1993; Anisko et al., 1994). The TDR indirectly determines the closed or intra-particle porosity without requiring particle density determinations. Closed porosity can introduce an estimation error of up to 10% of the total porosity fraction for substrates containing a significant proportion of closed-pore materials such as perlite (Waller and Harrison, 1991).

C. Air and Water Storage

Water retention is related to water potential (Hillel, 1980). By convention, a value of zero is given to pure water. Water potential of unsaturated soils is the partial result of gravitational and matrix potentials (Figure 4.1). Osmotic potential also develops due to solute accumulation at the external root boundary. Water potential is normally expressed in energy units (Joule or J). At constant temperature and pressure, the energy can be expressed relative to the volume (e.g., Joule by volume

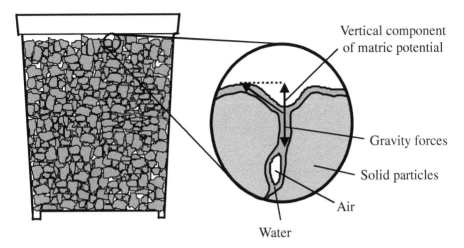

Figure 4.1 Suction and gravity forces exerted on a water molecule at container capacity.

unit), which is equivalent to pascal (Pa). Water potential can be expressed on a weight basis (m after simplification). The latter is convenient for describing water potentials at various locations in a pot.

The quantification of water and air stored in peat mixes requires water desorption curves as key characteristics for quality control and standardization. Sample preparation involves filling materials into cylinders (single or double ring), saturation and equilibration at −1 kPa prior to resaturation (Verdonck and Gabriëls, 1991), or saturation and drainage to container capacity three times to reach constant final volume (Bragg and Chambers, 1988; Caron et al., 1997). Targeted water potential is obtained in the laboratory using Tempe cells (Joyal et al., 1989) and sand or glass bead boxes (Topp and Zebchuck, 1979). Water potential can be determined directly in potted substrates that are in contact with a tension plate. Water release curves can be obtained directly in pots without a tension table by using growing plants or evaporation demand to dry the substrate to targeted water potentials (Allaire et al., 1996). Simultaneous measurements of volumetric water content and potential are inferred from TDR and tensiometric readings (Paquet et al., 1993). In Europe, the widely used CEN (EN-13041) methods have been implemented as part of voluntary or compulsory quality control procedures.

Water release curves are sensitive to handling, therefore, samples with bulk density similar to that observed in the greenhouse or nursery should be used for laboratory testing. The principle for determining the water release curve in the laboratory is similar to that in pots. A previously saturated soil core is gradually desaturated by applying either a pressure (Tempe cell) or a suction (sand box) to the core in order to establish an equilibrium pressure or tension; thereafter, water content is determined gravimetrically or by TDR. Because the curve follows a drying gradient, it is called a water release curve. A discrepancy occurs, however, between a curve obtained upon drying compared with that upon rewetting (Figure 4.2), because of pore constrictions (Hillel, 1980), structural changes (Michel, 1998), or changes in wettability (Michel, 1998). This is called hysteresis. Differences in pattern

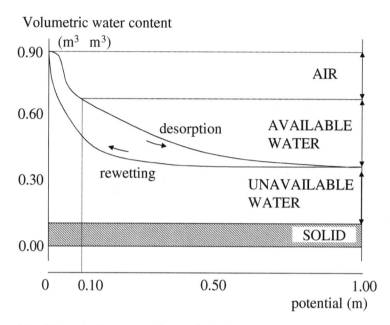

Figure 4.2   Water retention curves of a peat bark mix upon desorption or rewetting.

affect the measurement of water and air stored in substrates (Otten, 1994). Irrigation systems supplying water from above (e.g., mist, drip, or overhead) follow a drying curve; subirrigation systems (e.g., ebb and flow and capillary mats) operate on the wetting curve. It is important to define water sorption curves under wetting and drying conditions.

The quality of artificial mixes is often determined at two water potentials on the water release curve: container capacity (CC) and the lower limit of water availability. These parameters estimate air content after saturation and drainage, as well as the amount of water available for plant growth. Container capacity is water content at –1 kPa (–0.10 m) (White, 1965), which is the maximum amount of water retained in a substrate potted in a 20-cm high cylindrical container or a raised bed. The CC value varies with container geometry (Bilderback and Fonteno, 1987; Rivière, 1990), because water potential at CC is at thermodynamic equilibrium between suction and gravity forces (Figure 4.1).

Plant yield decreases as matrix potential in substrate is lowered below –10 kPa (De Boodt and Verdonck, 1972), although this limit varies with species, root distribution, salt accumulation and experimental conditions (Marfa et al., 1983; Röber and Hafez, 1982; Spoomer and Langhans, 1975; Goodman, 1983; Bernier et al., 1995; Caron et al., 1998). Higher matrix potentials result in increased growth rates of green plants, whereas lower water potentials may be maintained to induce flowering or hardiness, or restrict stem elongation (Morel et al., 1999). The difference in water content between total porosity and container capacity is air-filled porosity. The risk of root asphyxia is negligible if the difference exceeds 0.20 m³ m⁻³, but varies with species (Table 4.2).

Table 4.2 **Approximate Root Aeration Requirements of Selected Ornamental Plants, Expressed as Air-Filled Porosity (Matric Potential at −1 kPa)**

Aeration Requirements			
Very high	High	Intermediate	Low
Air-Filled Porosity (m³ m⁻³)			
0.20	0.10–0.20	0.05–0.10	0.02–0.05
Azalea	Antirrhinum	Camellia	Carnation
Orchid (epiphytic)	Begonia	Chrysanthemum	Conifer
	Daphne	Gladiolus	Geranium
	Erica	Hydrangea	Ivy
	Foliage plants	Lily	Palm
	Gardinia	Poinsettia	Rose
	Gloxinia		Stock
	Orchid (terrestrial)		Stresitzia
	Podocarpus		Turf
	Rhododendron		
	Saintpaulia		

Source: From Johnson, P. 1968. *Horticultural and Agricultural Uses of Sawdust and Soil Amendments*. Paul Johnson, National City, California.

Except for hastening the flowering or to restrict growth, any hydric stress should be avoided (Shackel et al., 1997). The difference between CC and −10 kPa (Figure 4.2) is the volume of available water (De Boodt and Verdonck, 1972). The larger the volume, the higher the height of water column that can be applied at each irrigation, thus decreasing watering frequency. For overhead and drip irrigation, however, water is applied periodically and rapidly, and substrates must store large amounts of water if watering frequency is small. Although the volume of stored water can be as high as 0.50 to 0.60 m³/m⁻³ after irrigation and drainage, part of this water is unavailable to plants. Watering is required at early signs of a decline in xylem water potential. In practice, 20 to 30% of the substrate volume is occupied by available water, which provides water storage for 2 to 3 days to meet plants' needs. De Boodt and Verdonck (1972) partitioned the −1 to −10 kPa range into easily available water (−1 to −5 kPa) and water buffer capacity (−5 to −10 kPa). The −1 to −5 kPa range is the water reserve that allows maximum transpiration. Water buffer capacity is the water reserve used when sudden heat waves result in intense transpiration.

D. Air Exchange

Gas renewal must take place rapidly in substrates for the acquisition of oxygen and the release of CO_2, C_2H_2, and other gases by living plants and microbes. Air-filled porosity is the most widely used aeration index because it is closely related to gas diffusivity (King and Smith, 1987; Gislerod, 1982) and easy to standardize (Bragg, 1997). Although air-filled porosity indicates the proportion of large pores filled with air, it is not indicated whether pores are connected or not. Gas exchange is controlled by gas diffusivity, D_s, which depends on gas diffusivity in the air (D_o), air-filled porosity (θ_a), and the pore effectiveness coefficient (γ), as follows (King and Smith, 1987):

$$\frac{D_s}{D_o} = \gamma\theta_a \qquad (4.3)$$

A high diffusivity value is a desirable property for peat substrates used as growing media (Allaire et al., 1996). Using large compact fragments such as bark in peat-base mixes, gas diffusivity may be seriously impeded even when air-filled porosity remains constant, because of a significant drop in pore connectivity (Nkongolo and Caron, 1999). Pore tortuosity is related to the geometry of media, characterizing pore connectivity or specific size efficiency. Tortuosity can be calculated from gas, solute, water, acoustical, or electrical properties. Pore tortuosity values estimated from water flow and release curves were found to be more closely related to plant growth than air-filled porosity alone (Allaire et al., 1996; Nkongolo, 1996; Caron and Nkongolo, 1999; Caron et al. 2000) and vary widely among substrates (unpublished data).

The combination of air-filled porosity and pore efficiency, as in Equation 4.3, produces a gas diffusion index (D_s) for substrate quality assessment. Gas diffusivity D_s can be measured under transient-state conditions from the shape of the time-concentration curve (Rolston, 1986). This method is limited to substrate-filled cores and requires a gas chromatograph. For substrates in pots of various geometries and containing actively growing plants, unbiased estimates of the true gas diffusivity were obtained from saturated hydraulic conductivity and water release characteristics (Nkongolo, 1996; Allaire et al., 1996). Gas diffusivity is not a measure of flux, however, which also depends on the gas concentration gradient and source-sink relationships involving roots and microbes. Oxygen flux to the root, when it is determined from oxygen diffusion rate (ODR) by using platinum electrodes, is also a useful index (Gislerod, 1982; Bunt, 1991; Paul and Lee, 1976).

E. Hydraulic Conductivity

Saturated hydraulic conductivity (K_s) is sometimes used as an index of substrate quality (Mustin, 1987) and for characterization (Marfa, 1998; Allaire et al., 1994, 1996). The K_s represents the resistance of a substrate to saturated water flow when a constant water height is maintained at the top of a soil column. In artificial mixes, K_s varies between 0.01 and 0.50 cm s^{-1}, and a value of 0.08 cm s^{-1} is used as a guide for good-quality substrates (Mustin, 1987). The K_s is also used to evaluate unsaturated hydraulic conductivity (Marfa, 1998; Wallach et al., 1992) and assess gas diffusivity.

The K_s is determined in cylinders of variable heights and diameters, and is very sensitive to substrate compaction. Therefore, *in situ* methods are preferred for potted media. A pressure infiltrometer is introduced into the pot (Caron et al., 1997; Allaire et al., 1994). Care must be taken to avoid bypass flow along the pot walls. Bentonite can be applied to avoid such flow. Low tension (–0.01 kPa potential) infiltrometers can be used, as far as a pulse of water is applied (Caron et al., 2002). For water flow measurements performed in pot, the outflow configuration is markedly different from that of straight cylinders, thus precluding the use of the one-dimensional Darcy's law. Instead, a flow reduction factor is calculated by solving the three-dimensional Laplace equation (Allaire et al., 1994; Caron et al., 1997). The flow reduction factor applied to K_s measurements transforms K_s estimates obtained in the pot into K_s values equivalent to a straight cylinder. This approach is statistically equivalent to the measurement of K_s in open cylinders (Allaire et al., 1994).

The rate of capillary rise in the substrate is a determinant for managing irrigation in ebb-and-flow or capillary mat systems (Otten, 1994; Marfa, 1998). Capillary rise depends on unsaturated hydraulic conductivity, which is a measure of the apparent velocity of water as macropores drain out or are filled with water. Flow rate depends largely on pores involved in water transport, and declines sharply as potential or volumetric water content drops. The relationship between ψ and K down to -2 and -10 kPa is described by Gardner's equation as follows (Brandyk et al., 1989; Tardif and Caron, 1993):

$$k(\psi) = K_s e^{\alpha\psi} \tag{4.4}$$

The rate of decrease, linked to α, has been reported to vary between 9 and 18 m^{-1} (Tardif and Caron, 1993), and 7 and 9 m^{-1} (Caron et al., 1998). The $K(\psi) - \psi$ relationship is highly hysteretic, varying along drainage or rewetting cycles (Caron et al., 1997). To date, no interpretation of unsaturated hydraulic conductivity values has been made (Caron et al., 1997; Wever, 1995; Marfa, 1998).

Methods for measuring unsaturated hydraulic conductivity are classified as steady-state or transient-state. Steady-state methods (Klute and Dirksen, 1986) are used for water potentials down to about -1 and -6 kPa. Below these values, too much time is required to establish steady-state conditions, and fluxes become very small in relation to experimental error. Transient-state methods use a combination of TDR probes and tensiometers at two heights during drainage or rewetting to estimate the $K(\psi) - \psi$ curve (Caron et al., 1997) and the continuity equation. Starting from a nearly saturated state, samples in pots or filled into cylinders are affixed to a tension plate to allow drainage. The samples are equipped with tensiometers at two vertical positions for measuring the hydraulic gradients at different times. A TDR probe, at full cylinder length, monitors changes in water content during drainage. A plastic sheet prevents flux through the surface. The $K(\psi)$ is calculated from changes in the water content and hydraulic gradient (Hillel, 1980).

F. Wettability

The hydrophobic character of organic matter after extreme drying is depicted by the angle formed by a water drop deposited on a dried, flat surface. Hydrophobicity is pronounced in peat compared with bark (Michel, 1998), but may disappear during composting. Well-decomposed peat acquires hydrophobicity as the water potential drops below -100 Pa, presumably due to the rotation of polar groups presenting hydrophobic ends from internal to external surfaces (Michel, 1998). Using capillary rise methods with various liquids, Michel (1998) showed that the energy of adsorption upon drying was linked mainly to nonpolar groups (essentially Van der Walls forces). Water sorption upon rewetting involved mainly Lewis-type (acid-base) links. The loss of acid-base-type links during drying could be caused by the protonation of carboxylic groups in *Sphagnum* mosses as surface acidity increases upon drying, thus minimizing Lewis-type interactions.

The hydrophobic behavior of organic substrates has a negative impact on irrigation management as both the time required for rewetting and the volume of water retained are reduced. Water redistribution within substrates is poorer because an unstable wetting front is formed and pockets remain permanently dry. A dried substrate takes more time to rewet. The more localized the irrigation, the longer is the rewetting period. The most frequently used method to measure wettability is the sessile drop (Michel et al., 1997), where the contact angle is measured using a goniometer. The drawbacks of this method include difficulty obtaining a smooth, planar, isotropic and flat surface. Further measurements can only be taken on air-dried samples. A second method, based on capillary rise, may be successfully used with wetting angles less than 90° (Michel et al., 1997). The wetting angles can be determined at different initial water contents. Contact angles can be measured during the drying process.

G. Cation Exchange Capacity and Peat Acidity

Substrates containing carboxylic and phenolic groups have the capacity to sorb cations (Puustjärvi, 1982). Cation exchange capacity (CEC) is expressed in $cmol_c$ kg^{-1} or on a volume basis. The CEC is pH-dependent, because carboxylic and phenolic groups can accept protons under acidic conditions. The CEC should be reported according to pH. In fibric *Sphagnum* peat, most of available exchange sites are occupied by H^+ ions. Considerable exchangeable acidity can be released into solution and must be taken into consideration when planning fertilization. A high CEC reduces the risk of a sudden variation in salt concentration in the liquid phase.

Substrate pH is determined in a suspension. Most often, 1:1, 1:2, or 1:5 mixtures are prepared with pure water, but sometimes with 1 M KCl or 0.01 M $CaCl_2$ solutions. Plant species have different pH requirements (Table 4.3). *Sphagnum* peat has a naturally acidic pH between 3.5 and 4.5. Thus, most peat-base substrates require liming, usually with horticultural grade dolomitic and calcitic limestone at 6 to 12 kg of limestone per cubic meter of substrate, which enhances the pH to about 5.4. The SMP (Shoemaker–McLean–Pratt) buffer-pH method can be used to estimate lime requirements (Van Lierop, 1983) (Table 4.4), but adjustments are needed because final pH is affected by lime efficiency, peat properties, time, and exchangeable acidity.

H. Nutrient Availability

Most elements are readily available in their exchangeable or soluble forms. Precipitation may occur, however, thereby reducing nutrient availability for some time. In particular, calcium solubility depends on carbonate (CO_3^{2-}), sulfate (SO_4^{2-}) and phosphate ($H_2PO_4^-$) ion activities in solution. The formation of Ca complexes

Table 4.3 Optimum pH (1 N KCl) Ranges for Plant Species

Species	pH Range
Camellia – Rhododendron	4.0–5.0
Picea abies	4.5–5.0
Betula – Magnolia – Quercus – Salix – Virbunum	5.0–6.0
Abies – Deutzia – Philadelphus – Syringa – Weigela	Around 6.0
Acer – Fagus – Pinus austriaca – Pinus montana	6.0–7.0
Aesculus – Crataegus – Forsythia – Fraxinus – Larix – Malus – Pouplus – Robinia – Sambucus – Tamarix – Tilia – Ulmus	6.0–8.0
Azalea	3.5–5.5
Chrysanthemum	5.5–8.0
Cinerea	6.0–8.5
Cyclamen	5.0–7.0
Hortensia	4.0–6.8
Kalanchoë	5.5–7.5
Lilium	5.0–7.0
Pelargonium	5.5–7.5
Primaveria	5.5–7.5
Saintpaulia	6.0–7.0

Source: From Lemaire, F. et al. 1989. *Pot and Container Culture: Agronomic Principles and Applications* (in French). Institut National de la Recherche Agronomique, Paris. With permission.

Table 4.4 Peat Lime Requirements as a Function of SMP Buffer pH

Buffer pH	Limestone Requirements[a] (kg m³)
6.0	0.0
5.8	1.1
5.6	2.3
5.4	3.5
5.2	4.6
5.0	5.8
4.8	6.9
4.6	8.1
4.4	9.2
4.2	10.4
4.0	11.5

[a] $CaCO_3$ assumed 100% efficient.

Source: From Van Lierop, W. 1983. *Can. J. Soil Sci.*, 63:411–423.

is pH and concentration dependent. Precipitation is reversible depending on ion concentrations and pH. Under acidic conditions, phosphate may precipitate with iron, and consequently, some peat high in iron fix phosphorus (Lucas, 1982). Under alkaline conditions, phosphate tends to precipitate with calcium (Lemaire et al., 1989). Relative availability of ions in organic soils as a function of pH is summarized in Figure 4.3.

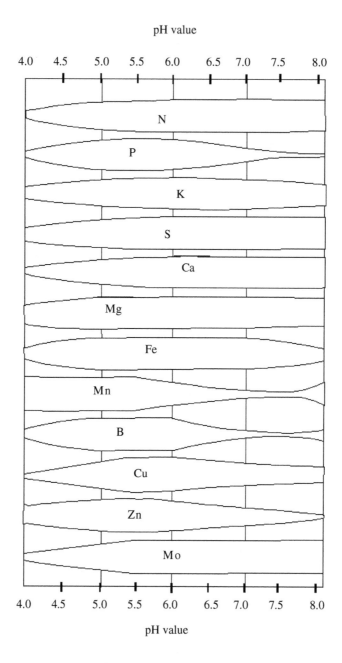

Figure 4.3 Nutrient availability in organic soils. (Adapted from Lucas, R. 1982. Organic soils: Formation, distribution, physical and chemical properties and management for crop production. Research Rep. No. 435. Michigan State University, East Lansing, Michigan. With permission.)

Table 4.5   Guidelines for Chemical Composition of Solution Extracts[a] in Artificial Mixes

Parameter	Adequate value
Electrical conductivity	<1.0–1.5 dS m⁻¹ for seeding and transplanting and 0.75–3.5 dS m⁻¹ thereafter
pH	5.5–6.5 (4.5–5.5 for acid-loving plants)
Ionic species	μg L⁻¹
NO_3^-	<100
Ca^2	<400
Mg^{2+}	<100
$H_2PO_4^-$	<50
K^+	<200
B	<0.5
Cu^{2+}	<0.5
Fe^{3+}	<3.0
Mn^{2+}	<3.0
Zn^{2+}	<3.0
Na^+	<50

[a] Saturation medium extract.

Various methods are used to characterize nutrient concentrations and quantities. They can be divided into three types: suspension, saturated extraction, and pour-through methods. The end result must be interpreted using a chart specific to the method used. A detailed description of chemical analysis of lightweight media, along with some interpretation guidelines, is provided in Bunt (1988). The optimum nutrient composition of a substrate based on the saturated medium extract method is presented in Table 4.5.

I.   Substrate Mineralization and Reorganization

Organic materials are susceptible to microbial decay, leading to the mineralization of organic N, P, and S. Because of limited control on the kinetics of those processes, biologically stable substrates are preferred by growers. In stable substrates, nutrients are supplied as liquid, solid, or slow-release fertilizers, which are added during manufacturing or at regular intervals with the irrigation water, or by top-dressing.

Peat materials evolve slowly, even when their environment is modified by fertilization. No general relationship exists between peat stability and the C/N ratio, which can be as high as 60 for fibric peat and around 20–25 for hemic peat materials. *Sphagnum* residues are difficult to decompose due to lignin-like substances called sphagnols, which slow down microbial decay (Puutsjärvi, 1982).

Despite their relative stability, peat materials follow two stages of settling: a rapid settling after potting, wetting, and drying; and a phase of gradual settling resulting from decomposition of large particles, shrinkage, and particle reorientation (Nash and Laiche, 1981; Bures et al., 1993). The first stage is primarily physical, and the second is biological. During the initial short-term structural loss, peat structure collapses during drainage cycles. The ratio of initial to final volumes after the first two or three wetting cycles is about 1.5:1, and depends on many factors such as peat type, applied pressure, initial water content, and irrigation system. A large volume of water applied in a short period results in a marked compaction (Bragg and Chambers, 1988). The second phase of compaction brings the ratio to about 2:1 within 3 to 6 months and can be deleterious for long-term cultivation. Part of the settling results from vermiculite, which is of limited use for long-term cultivation. Because composts have a tendency to lose their original structures more rapidly than peat, less than 20% compost should be added to the mix for long-term use.

Compaction and settling result in a loss of air-filled porosity and an increase in water availability. Such loss of air-filled porosity may result in decreased aeration (Puustjärvi, 1976); however, peat substrates can show a concomitant increase in pore efficiency compensating the decrease in pore size, so that gas diffusivity remains unchanged (Allaire-Leung et al., 1999). Nash and Laiche (1981) also observed a significant improvement in saturated hydraulic conductivity of peat substrates with time.

III. PEAT TYPES AND SUBSTRATE QUALITY

Many primary materials can be used to make growing media. Peat materials vary with botanical origin, granulometry, and decomposition level. The *Sphagnum*, hypnum, herbaceous (*Carex* and *Phragmites*), and woody peat materials are distinguished from their botanical compositions (Rivière, 1992). Native *Sphagnum* peat contains a large number of macropores compared with sapric *Carex* and woody peats (Paivanen, 1982). *Sphagnum* peat is the preferred peat material for making substrates.

The proportion of water and air stored at a water potential of −1 kPa varies widely among substrates (Figure 4.4). Fibric blonde peat, frozen sapric peat and hydrophilic rockwool show high levels of stored water that is mainly retained in a mesh of fibers (Valat, 1989; Puustjärvi, 1976). Specialized cell structures, called hydrocysts, which are 14 to 36 μm in diameter depending on *Sphagnum* species, retain large amounts of water between −10 and −100 kPa. Generally, hemic brown, fibric blonde, and frozen black sapric peat materials are used for their water retention properties. Structuring components are added when needed to improve aeration.

Fibric blonde, hemic brown, or sapric black peat materials refer to their decomposition degree as evaluated on the Von Post scale (Von Post and Granlund, 1926). As the degree of decomposition increases, air-filled porosity and available water decrease, despite an overall increase in total water stored at a potential of −1 kPa. This diminution of stored air is usually linked to an increase in particles smaller than 1 mm (Puustjärvi, 1976) or 0.5 mm (Verdonck and Gabriëls, 1991), as shown in Figure 4.5. Water retention varies with the *Sphagnum* species. Hence, for fibric blonde peat (H1 to H3 on the Von Post scale), optimum water retention is obtained with *Sphagnum papillosum* (Puustjärvi, 1976).

The potential use of peat is linked to peat fiber size as measured by the rubbed fiber test giving the volume percentage of fibers left on a 0.15-mm sieve (Parent and Caron, 1993), or determined by wet (Dinel and Lévesque, 1978) or dry (Caron et al., 1997) sieving procedures. Decrease in fiber size leads to marked changes in physical

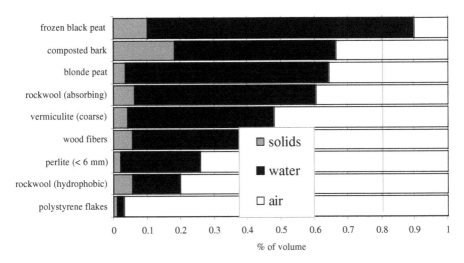

Figure 4.4   Proportion of air and water stored in different media.

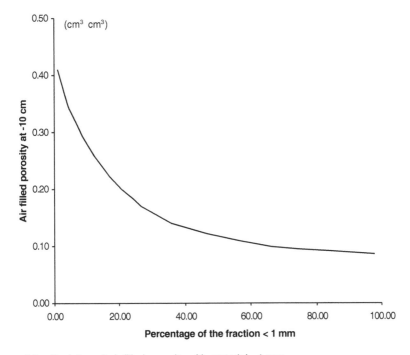

Figure 4.5   Evolution of air-filled porosity with material <1 mm.

properties. Although a higher proportion of long fibers is generally found in fibric blonde than in hemic brown peat materials, peat harvesting and processing methods can modify the original peat characteristics.

Fibric blonde peat has high air- (>0.20 m^3 m^{-3}) and water- (about 0.30 m^3 m^{-3}) filled porosity (Rivière, 1992). Consequently, fibric blonde peat is of prime value

Table 4.6   Total Porosity (TP), Air-filled Porosity (θ_a), Gas Relative Diffusivity (D_s/D_o), and Mean Weight Diameter (MWD) of a Blonde Peat, a Brown Peat, and a Mixture of Both

Peat Type	TP (m^3 m^{-3})	θ_a (m^3 m^{-3})	D_s/D_o (m^2 s)/(m^2 s^{-1})	MWD[a] (mm)
Fibric blonde	0.96 a[b]	0.23 a	0.0042 b	2.2 b
Fibric blonde-brown	0.95 a	0.16 b	0.0045 b	1.9 b
Hemic brown	0.87 b	0.17 b	0.0066 a	2.9 a
LSD (0.05)[c]	0.017	0.040	0.0016	0.25

[a] Mean weight diameter according to Kemper, W.D. and Roseneau, R.C. (1986).
[b] Numbers with different letters differ significantly at the 0.05 probability level, using a protected least significant difference (LSD) test (Snedecor, G.W. and Cochrane, W.G., *Statistical Methods*, Iowa State University Press, Ames, Iowa, 1989).

because the risk of asphyxia and the need for expensive structuring agents are minimized. Hemic brown peats (H4–H5) may lack adequate aeration and contain a higher proportion of water unavailable to plants; however, these general features depend on particle size distribution. For example, when fibric blonde peat contains few long fibers, they lack aeration and shrink considerably upon drying (Rivière, 1992). Table 4.6 presents a hemic brown peat, a fibric blonde peat and a mixture of both, along with an index characterizing their respective particle size distributions and mean weight diameters. A hemic brown peat containing no particles smaller than 0.25 mm but some large fragments may show a higher gas diffusivity than a fibric blonde peat, despite lower air-filled porosity and total porosity. Particle size distribution is thus as important a peat quality index as degree of decomposition. Fibric herbaceous peat may have adequate aeration characteristics but often lack sufficient available water (Rivière, 1992). Other peat types (woody peat, sapric herbaceous peat) are used to make potting soil or as soil amendments for landscaping.

Attempts have been made to improve water retention of peat using hydrogels. The effect of these hydrogels is very variable, sometimes improving water retention, sometimes leaving it unaffected (Rivière et al., 1996). The result appears to depend not only on the rate of irrigation (Rivière et al., 1996), but also on time elapsed after irrigation. The effectiveness of hydrogels usually disappears within a normal growth period (Jobin, 2000). Because the plants are purchased after 8 to 16 weeks of growth, hydrogels are rarely used in peat substrates.

Peat wettability varies with peat type and is sometimes problematic. *Sphagnum* peats are much more wettable than herbaceous or woody peats (Valat et al., 1991). Contact angles near 0° for wettable surfaces exceed 90° for most air-dried peat materials (Valat et al., 1991), a feature characteristic of hydrophobic surfaces. For air-dried materials, the contact angle reaches 110° for *Sphagnum* peat, 116.8° for herbaceous peat, and 122.1° for woody peat (Valat et al., 1991). Results are sometimes contradictory. As shown by Michel (1998), who conducted contact angle measurements at different hydration levels, the hydrophobic character of hemic peat materials were more pronounced in *Sphagnum* than in herbaceous peat materials (Table 4.7). Michel (1998) suggested that such differences might be attributed to

Table 4.7   Influence of Water Potential and Peat Type on Wetting Angle

Water Potential	Wetting Angle (Degree)			
	Fibric Blonde *Sphagnum* peat	Hemic Brown *Sphagnum* Peat	Fibric Blonde Herbaceous Peat	Hemic Brown Herbaceous Peat
−32 kPa	66.2	690.0	69.2	70.0
−100 kPa	84.5	86.2	81.4	81.7
−3 MPa	93.6	96.8	97.2	87.1
−100 MPa	104.5	106.4	106.5	88.4

Source: From Michel, J.C. 1998. Studies on wettability of organic media used as culture media (in French). Ph.D. thesis, Ecole Nationale Supérieure Agronomique de Rennes, Rennes, France. With permission.

functional groups and to a considerable increase in mineral fraction with decomposition of herbaceous peat. Wettability decreased with degree of hydration; *Sphagnum* and herbaceous peat materials became clearly hydrophobic as the material was allowed to dry to less than −3 MPa (Michel, 1998) (Table 4.7).

Frozen peat tends to be slightly less hydrophobic than nontreated peat (Van Dijk and Boekel, 1965), but the addition of a wetting agent or of a clay material appears to be a more efficient way than freezing to improve wettability (Michel, 1998). With a wetting agent or clay pellets, the hydrophobic behavior did not appear to limit the use of peat in substrates (Michel, 1998). Problems with hydrophobic peat may be partially solved by using surfactants. Most surfactants are organic chains with hydrophobic sites as aliphatic carbon chains, as well as hydrophilic anionic, cationic, or non-ionic sites; the non-ionic sites show lesser toxicity. These products, which modify water surface tension, facilitate peat rewetting (Sheldrake and Metkin, 1971; Michel et al., 1997); however, the modification of water surface tension properties interferes with water retention (Rivière et al., 1996). Water retention in treated substrates sometimes decreases significantly at high water potentials that are close to saturation. The use of surfactants in peat substrates can be limited by their toxicity to plants. When peat is used in combination with sand or clay, wettability problems are often diminished (Michel, 1998).

Peat materials are naturally low in salts and nutrients (Gonzalez, 1981, 1991). Although peat materials have a high CEC based on weight (96 to 136 $cmol_c$ kg^{-1} according to Puustjärvi, 1969), their bulk densities are low. Peat substrates therefore have low cation retention capacity. Phosphates are easily leached because peats contain little native aluminium, iron or calcium (Lucas, 1982). On the other hand, peat substrates can fix large amounts of micronutrients through chelation. Thus, special formulations enriched in micronutrients must account for chelation. The CEC is not taken into account when selecting peat materials because of small differences among peat types.

Because of several advantages regarding air and water storage, low salt and fertility levels, availability, stability, and low cost, *Sphagnum* peat is widely used for making artificial mixes. Fibric blonde peat appears most suitable for making artificial mixes, although hemic brown peat is of interest if adequately sieved. Brown, black, herbaceous, and woody peat materials are used mainly as soil amendment for landscaping, transplanting, and soil improvement.

Table 4.8 Air-Filled Porosity of Two Peat Types as a
Function of Applied Pressure (Verdonck and
Gabriëls, 1991)

Applied Pressure (Pa)	Air-Filled Porosity ($m^3\ m^{-3}$)	Available Water ($m^3\ m^{-3}$)
Sapric Black Peat		
0	0.17	0.25
12.5	0.06	0.29
25.0	0.02	0.32
Fibric Blonde *Sphagnum* Peat		
0	0.13	0.40
12.5	0.11	0.41
25.0	0.09	0.34

IV. STABILITY AND RECYCLING OF PEAT SUBSTRATES

A. Stability of Peat Substrates

Substrate quality is affected by peat compaction during manufacturing or by settling and decomposition after potting. Verdonck and Gabriëls (1991) reported marked differences in aeration with applied pressure among peat types (Table 4.8). Heiskanen et al. (1996) reported significant effects of filling, watering, and peat type on air-filled porosity. Peat decomposition occurs following potting, transplanting, irrigation, and fertilization, with corresponding losses in air-filled porosity (Rivière, 1992). With an increase in root growth, peat drying is promoted, resulting in peat shrinkage. Fibric blonde peat is less prone to shrink than hemic brown or sapric black peat, because of greater mechanical resistance (Verdonck and Gabriëls, 1991). Puustjärvi (1982) observed that peat susceptibility to decomposition was greatly increased by shrubs, which led to a considerable loss of structure over time relative to pure *Sphagnum* peat; the latter kept its original structure for at least 15 months of cultivation. Pure *Sphagnum* substrates were resistant to decomposition for over 7 years (Puustjärvi, 1982). The *Sphagnum* species showed variable degrees of mineralization from a 4.7% dry-matter loss for *Sphagnum fuscum* to a 10.3% dry-matter loss for *Sphagnum cuspidatum* after 13 months at 40°C. Prasad and O'Shea (1997) confirmed Puustjärvi's observations and found that long fibers showed the smallest volume loss over time, and that volume loss was smaller for peat than wood bark.

B. Recycling of Peat Substrates

The recycling of peat substrates is usually not an environmental issue, because peat substrates are part of the marketed product. Some studies have addressed the question of reusing peat bags for vegetable production (Woods and Maher, 1971; Baevre and Guttermsen, 1984). Baevre and Guttermsen (1984), using a hemic peat, observed comparable yields after two years of use and a slight diminution in yield after the third year, without any sterilization. Woods and Maher (1971) obtained comparable yields for two successive years, but emphasized the need for sterilization. For the peat basin culture, Puustjärvi (1982) presented data from rose beds showing very little change in bulk density over time for up to 7 years. He pointed out that these results were obtained in a pure *Sphagnum* bed, and that the presence of shrubs led to a significant increase in bulk density. Any loss in air-filled and total porosities could have a major impact on plant growth by decreasing aeration. In contrast, Allaire-Leung et al. (1999) showed that aeration remained unchanged during plant growth despite a decrease in total and air-filled porosities, because of an increase in pore efficiency (they reported a decrease with time of pore tortuosity).

C. Peat Extraction and Processing Technologies

Peat used in artificial mixes is usually harvested using the sod peat or the milled peat methods (Puustjärvi, 1973). In the sod peat method, the machine excavates peat in lumps that are piled in loose heaps in the field for drying. The peat material freezes during the winter, thus improving water retention and wettability of the black sapric peat (Michel, 1998; Van Dijk and Boekel, 1965), and the breaking up tissue remnants (Swinnerton and Ripley, 1948). The lumps are shipped from the stockpile to the factory where they are crushed, sieved and processed. This technology is still widely used in Europe, but has essentially disappeared in North America. The technique is unsuitable when there are stumps in the bog (Puustjärvi, 1973).

The milled peat technique, originally developed in the former USSR for harvesting fuel peat, is now widely used, particularly in North America, for harvesting horticultural peat. The surface of the bog is disk-harrowed or milled with a rotating drum with extended spikes (Lucas, 1982). This procedure loosens the top surface and hastens the drying process. Once air-dried (additional harrowing may be performed if necessary), the loose peat is ridged with a V-shaped or lateral windrower and picked up by a tractor equipped with a lift scoop or a conveyor. Alternatively, a pneumatic harvesting machine can be used to collect air-dried peat without ridging. The collected peat is stored in piles before crushing and sieving.

Generally, the sod technique produces coarser peat fragments with a more stable structure compared with the peat harvested using the milled peat method (Puutsjärvi, 1976; McNeill et al., 1983). The structure of sod peat is so loose (high macroporosity), however, that it is only suited for the production of species with high aeration requirements, such as orchids (Table 4.2), or for mushroom production. Mixes manufactured with sod peat are characterized by excessively rapid drainage and leaching. Sod peat is therefore used as a structuring agent in mixes. A mixture of milled hemic brown and milled fibric blonde peat materials is commonly used in peat mixes.

The sieving and grinding technology involved in peat processing is an additional source of variability in the end product. The selection of appropriate sieving and crushing sizes may have a significant effect on subsequent crop performance. The duration of sieving is critical. Puustjärvi (1973) observed that prolonged sieving resulted in the breakdown of coarser fragments into fragments of less than 1 mm. Because the finer fraction is associated with a diminution in air-filled porosity (Figure 4.5), the duration of sieving should be limited. It is common practice to crush the coarse fraction to obtain fractions smaller than 8–12 mm or fractions between 8 and 30, or 50 mm to improve aeration. Nkongolo (1996) noted the negative effect of excessively coarse bark particles on the yield of nursery and greenhouse species. Preliminary results by Caron et al. (2002) indicated that this also applies to peat, although to a lesser extent.

The mixing step is a third source of variability. More handling causes more particle breakdown, particularly when the peat is too dry. Voids of structuring materials are filled by the smaller particles during mixing (Lemaire et al., 1989). Although it is not generally the case with peat and bark (Rivière, 1988), this becomes increasingly important when a component of high porosity (peat) is mixed with a low-porosity material (sand), particularly under dry conditions. The duration of mixing is critical (Rivière, 1996). The probability that fine particles occupy voids in a position of lower free energy located below larger particles increases with the duration and intensity of mixing (Bures et al., 1993). Both processes occur during mechanical potting performed at the nursery, if intensive mixing is carried out for prolonged periods (McNeill et al., 1983).

Additional handling during packing can result in the breakage of larger fragments, but only few references exist on that topic. A final factor to consider is the compaction during potting. As emphasized previously, the structure of peat substrates is highly sensitive to handling. Initial water content, pressure applied at the time of potting, and watering performed at the time of potting may have a marked impact on the potted product (Heiskanen et al., 1996, and Table 4.8). To avoid excessive compaction, substrates should be potted with care and should not be too wet. Compaction should be firm enough to ensure adequate contact between the seedling or the seed and the substrate. The use of a screw conveyer for automated potting appears damaging for peat substrates (Heiskanen et al., 1996). Following transplantation, irrigation should be performed at low rates in order to avoid the migration of particles and the filling of the empty pore space by fine particles.

V. CONCLUSION

The physical quality of peat substrates is assessed by measuring air and water storage as well as exchange properties. Chemical quality assessments involve pH, salt, and fertility levels in substrate-saturated extracts. Normalized methods exist or are being developed to measure quality attributes, but so far only chemical and air and water storage properties are being used. Gas exchange properties need to be determined because they regulate growth dynamics in potted media and are more closely correlated to plant growth than gas storage properties. Fibric and hemic *Sphagnum* and herbaceous peat materials usually meet the required quality standards for substrates, amended or not. The harvesting, processing, packaging, and handling procedures affect the quality of the end product.

ACKNOWLEDGMENTS

The authors thank France Chabot and Carmen Bilodeau for the typesetting, Luc Bidel for the figures, Martin Bolinder for assisting translations, and Nicole De Rouin for reviewing the manuscript.

REFERENCES

Allaire, S., Caron, J., and Gallichand, J. 1994. Measuring the saturated hydraulic conductivity of peat substrates in nursery containers. *Can. J. Soil Sci.,*74:431–437.

Allaire, S. et al. 1996. Air-filled porosity, gas relative diffusivity and tortuosity: Indices of *Prunus* x *cistena* growth in peat substrate. *J. Am. Soc. Hortic. Sci.,* 121:236–242 (corrigenda 121:592).

Allaire-Leung, S.E., Caron, J., and Parent, L.E. 1999. Improvement of physical and chemical properties of four ornamental substrates during plant growth. *Can. J. Soil Sci.,* 79:137–139.

Anisko, T., NeSmith, D.S., and Lindstrom O.M. 1994. Time domain reflectrometry for measuring water content of organic growing media in containers. *Hortscience,* 29:1511–1513.

Baevre, O.A. and Guttermsen, G. 1984. Reuse of peat bags tomatoes and cucumbers. *Plant Soil,* 77:207–214.

Beardsell, D.V., Nichols, D.G., and Jones, D.L. 1979. Physical properties of nursery potting mixtures. *Scientia Hortic.,* 11:1–8.

Bernier, P.Y., Stewart, J.D., and Gonzalez, A. 1995. Effects of the physical properties of *Sphagnum* peat on water stress in containerized *Picea mariana* seedlings under simulated field conditions. *Scand. J. For. Res.,* 10:184–189.

Bilderback, T.E. and Fonteno, W.C. 1987. Effects of container geometry and media physical properties on air and water volumes in containers. *J. Envir. Hortic.,* 5:180–182.

Bragg, N.C. 1997. The use of physical determinations in horticultural science, in *Peat in Horticulture, Proc. Int. Peat Conf.* Schmilewski, G., Ed., Amsterdam, The Netherlands, Nov. 2–7, 39–42, Paino Porras Oy, Jyraskylå, Finland.

Bragg, N.C. and Chambers, B.J. 1988. Interpretation and advisory applications of compost air-filled porosity measurements. *Acta Hortic.,* 221:35–44.

Brandyk, T., Romanowicz, R., and Skapski, K. 1989. Determination of the exponential unsaturated hydraulic conductivity function by parameter identification. *J. Soil Sci.,* 40:261–267.

Bunt, A.C. 1984. Physical properties of mixtures of peat and minerals of different particle size and bulk density for potting substrates. *Acta Hortic.,* 150:143–153.

Bunt, A.C. 1988. *Media and Mixes for Container Grown Plants.* Urwin Hyman, London, England.

Bunt, A.C. 1991. The relationship of oxygen diffusion rate to the air-filled porosity of potting substrates. *Acta Hortic.,* 294:215–224.

Bures, S. et al. 1993. Shrinkage, porosity characterization and computer simulation of horticultural substrates. *Acta Hortic.,* 342:229–234.

Caron, J., et al. 1997. Determination methods for physical properties of artificial media, peats and organic soils (in French). Publication V-9706 (ISBN 2–89457–139–9). Conseil des productions végétales du Québec Inc., Quebec, Canada.

Caron, J. and Nkongolo, V.K.N. 1999. Aeration in growing media: recent developments. *Acta Hortic.*, 481:545–551.

Caron, J. et al. 1998. Water availability in three artificial substrates during *Prunus* x *cystena* growth: Variable threshold values. *J. Amer. Soc. Hortic. Sci.*, 123:931–936.

Caron, J., Rivière, L.M., and Morel, P. 2000. Aeration in growing media containing large particle sizes. Abstracts of technical paper. *Int. Symp. Growing Media and Hydroponics*, Aug. 31– Sept. 6, Thessanoliki, Greece.

Caron, J. et al. 2002. Using time domain reflectometry to estimate hydraulic conductivity and air entry in growing media and sand. *Soil Sci. Soc. Am. J.*, 66:373–383.

Da Sylva, F.F. et al. 1998. Measuring water content of soil substitutes with time domain reflectometry. *J. Amer. Soc. Hortic. Sci.*, 123:734–737.

De Boodt, M. and Verdonck, O. 1972. The physical properties of the substrate in horticulture. *Acta Hortic.*, 26:37–44.

Dinel, H. and Lévesque, M. 1978. A simple technique for determining the grain-size distribution of peat materials in aqueous media (in French). *Can. J. Soil Sci.*, 56: 119–120.

Gabriels, R. 1995. Standardization of growing media analysis and evaluation: CEN/ISO/ISHS. *Chronica Hortic.* 35:5–6.

Gabriels, R. and Verdonck, O. 1991. Physical and chemical characterization of plant substrates: Towards a European standardization. *Acta Hortic.*, 294:249–259.

Gislerod, H.R. 1982. Physical conditions of propagation media and their influence on the rooting of cuttings, 1. Air content and oxygen diffusion at different moisture tensions. *Plant Soil*, 69:45–456.

Gonzalez, A. 1981. Brief description of the chemical and physico-chemical properties of peat materials used in containerised crops in Quebec (in French). Rep. LAU-X-48, Laurentian Forestry Research Centre, Sainte-Foy, Quebec, Canada.

Gonzalez, A. 1991. Effects of physical and physico-chemical properties of the substrates on growth of white spruce seedlings (in French), in *Proc. Symposium on Peat and Peatlands: Diversification and Innovation.* Quebec, Canada, Aug. 6–10, 1989, Overend, R.P. and Jeglum, J.K., Eds., 15–20.

Goodman, D. 1983. A portable tensiometer for the measurement of water tension in peat blocks. *J. Agr. Eng. Res.*, 28:179–182.

Gras, R. 1982. Some physical properties of horticultural substrates, 1. Porosity (in French). *Revue Horticole*, 230:51–55.

Heiskanen, J., Tervo, L., and Heinonen, J. 1996. Effects of mechanical container-filling methods on texture and water retention of peat growth media. *Scand. J. For. Res.*, 11:351–355.

Hidding, A. 1999. Standardization by CEN/TC 223 "Soil improvers and growing media" and its consequences, in *Peat in Horticulture, Proc. Int. Peat Conf.,* Amsterdam, The Netherlands, Nov. 1, Schmilewski, G. and Tonnis, W.J., Eds., 21–24.

Hillel, D. 1980. *Fundamentals of Soil Physics.* Academic Press, San Diego, CA.

Hood, G. 1998. Personal communication. Canadian Sphagnum Peat Moss Association, St. Albert, Alberta, Canada.

Jobin, P. 2000. Impact of incorporating two hydrophilic polymers on physical properties of three horticultural substrates (in French). M.Sc. thesis, Plant Science Department, Laval University, Quebec, Canada.

Johnson, P. 1968. *Horticultural and Agricultural Uses of Sawdust and Soil Amendments.* Paul Johnson, National City, CA.

Joyal, P., J. Blain, and Parent, L.E. 1989. Utilization of Tempe cells in the determination of physical properties of peat based substrates. *Acta Hortic.*, 238:63–65.

Kemper, W.D. and Roseneau, R.C. 1986. Aggregate stability and size distribution, in *Methods of Soil Analysis, Part 1*, 2nd ed. Page, A.L., Ed., Soil Sci. Soc. Am. Book Series 9, Madison, WI, 425–441.

King, J.A. and Smith, K.A. 1987. Gaseous diffusion through peat. *J. Soil Sci.*, 38:173–177.

Klute, A. and Dirksen, C. 1986. Hydraulic conductivity and diffusivity: Laboratory methods, in *Methods of Soil Analysis, Part 1*, 2nd ed. Page, A.L., Ed., Soil Sci. Soc. Am. Book Series 9, Madison, WI, 687–734

Lemaire, F. et al. 1989. *Pot and Container Culture: Agronomic Principles and Applications* (in French). Institut National de la Recherche Agronomique, Paris.

Lucas, R. 1982. Organic soils: Formation, distribution, physical and chemical properties and management for crop production. Research Rep. No. 435. Michigan State University, East Lansing, Michigan.

Marfa, O. 1998. Water management and saturated hydraulic conductivity of substrates. Measuring saturated hydraulic conductivity (in Spanish), in *IV Journadas del groupe de substratos de la SECH*, Sevilla, Spain, Sept. 22–24 (unpublished report).

Marfa, O. et al. 1983. Effect of different substrates and irrigation regimes on crop and plant-water relationships of *Asplenium nidus-avis* Hort and *Cyclamen persicum* Mill. *Acta Hortic.*, 150:337–348.

McNeill, D.B., Blom, T.J., and Hughes, J. 1983. Soilless mixes. Ontario Ministry of Agriculture and Food Publ. 83–021, Toronto, Ontario, Canada.

Michel, J.C. 1998. Studies on wettability of organic media used as culture media (in French). Ph.D. thesis, Ecole Nationale Supérieure Agronomique de Rennes, Rennes, France.

Michel, J.C. et al. 1997. Effects of wetting agents on the wettability of air-dried sphagnum peat, in *Peat in Horticulture. Proc. Int. Peat Conf.*, Schmilewski, G., Ed., Amsterdam, The Netherlands, Nov. 2–7, 74–80.

Morel, P., Rivière, L.M., and Julien, C. 1999. Measurement of representative volumes of substrates: which method, in which objective? Int. Soc. Hortic. Sci. XXV IHC Brussels. *Acta Hortic.*, 517:261–269.

Mustin, M. 1987. *Compost* (in French). Editions Francois Dubusc, Paris, France.

Nash, V.E. and Laiche, A.J. 1981. Changes in the characteristics of potting media with time. *Commun. Soil Sci. Plant Anal.*, 12(10):1011–1020.

Nkongolo, V.K.N. 1996. Pore space tortuosity : importance, evaluation and effects on vegetation (in French and English). Ph.D. thesis, Dept. Soil Sci. Agri-Food Engr., Laval University, Quebec, Canada.

Nkongolo, V.K.N. and Caron, J. 1999. Bark particle sizes and the modification of the physical properties of peat substrates. *Can. J. Soil Sci.*, 79:111–116.

Otten, W. 1994. Dynamics of nutrients for potted plants induced by flooded bench fertigation: Experiments and simulation. Ph.D. thesis, Horticultural Science Department, University of Wageningen, Wageningen, The Netherlands, ISBN 90–5485–304–2.

Paivanen, J. 1982. Main Physical Properties of Peat Soils. Peatlands and Their Utilization In Finland. Finnish Peatland Society, Helsinki, Finland, 33–36.

Paquet, J.M., Caron, J., and Banton, O. 1993. *In situ* determination of the water desorption characteristics of peat substrate. *Can. J. Soil Sci.*, 73:329–339.

Parent, L.E. and Caron, J. 1993. Physical properties of organic soils, in *Soil Sampling and Methods of Analysis*. Carter, M.R., Ed., Canadian Society of Soil Science and Lewis Publishers, Boca Raton, FL, 441–458.

Paul, J.L. and Lee, C.I. 1976. Relation between growth of chrysanthemums and aeration of various container media. *J. Amer. Soc. Hortic. Sci.*, 100:500–503.

Prasad, M. and O'Shea, J. 1997. Relative breakdown of peat and non-peat growing media. *Int. Symp. on Growing Media and Hydroponics,* Windsor, Ontario, Canada, May 19–26, Paper 18.

Puutsjärvi, V. 1969. Principles of fertilizing peat. *Peat Plant News,* 2:1–27.

Puutsjärvi, V. 1973. *Peat and Its Use in Horticulture.* Peat Research Institute, Turveteollisuusliittory, Helsinki, Finland.

Puutsjärvi, V. 1976. *Peat and Plant Yearbook 1973–1975.* Peat Research Institute, Helsinki, Finland.

Puutsjärvi, V. 1982. *Peat and Plant Yearbook 1981–1982.* Peat Research Institute, Helsinki, Finland.

Rivière, L.M. 1988. Physical behavior of mixes of materials used for making culture media (in French). *Compte Rendus de l'Académie d'Agriculture,* 74:53–61.

Rivière, L.M. 1990. Using the container capacity to guide irrigation of potted plants (in French). *Bulletin Technique d'Information, Ministère Français de l'Agriculture France,* 444:415–424.

Rivière, L.M. 1992. *Water Cycles of Substrate-Plant Systems in Soilless Culture* (in French). Université d'Angers, École nationale d'ingénieurs des travaux de l'horticulture et du paysage d'Angers, Angers, France, 141.

Rivière, L.M. 1996. Making of culture media by mixing materials (in French). *PHM Revue Horticole,* 373: 25–27.

Rivière, L.M., Bellon-Fontaine, M.N. and Banévitch, C. 1996. Modifications of hydric properties of culture media by adding tensio-active substances (in French). *PHM Revue Horticole,* 368:52–56.

Röber, R. and Hafez, M. 1982. The influence of different water supplies upon the growth of chrysanthemum. *Acta Hortic.,* 125:69–78.

Rolston, D.R. 1986. Gas diffusivity, in *Methods of Soil Analysis, Part 1,* 2nd ed. Page, A.L., Ed., Soil Sci. Soc. Am. Book Series 9, Madison, WI, 1094–1096.

Schmilewski, G. 1992. The possibilities and limits of substituting peat in growing media and for soil improvement, in *Proc. 9th Int. Peat Congr.,* Uppsala, Sweden, June, 22–26, International Peat Society, Jyuaskyla, Finland, 369–381.

Shackel, K.A. et al. 1997. Plant water status as an index of irrigation need in deciduous fruit trees. *HortTechnology,* 7:23–29.

Sheldrake, R. and Metkin O.A. 1971. Wetting agents for peat moss. *Acta Hortic.,* 18: 37–42.

Snedecor, G.W. and W.G. Cochran.1989. *Statistical Methods,* 8th ed. Iowa State University Press, Ames, Iowa.

Spoomer, L.A. and R.W. Langhans. 1975. The growth of *Chrysanthemum morifolium* Ramat. at high soil water contents: Effects of soil water and aeration. *Commun. Soil Sci. Plant Anal.,* 6:545–553.

Swinnerton, A.A. and Ripley, P.O. 1948. The use of peat in agricultural materials. Publ. No. 803, Canada Ministry of Agriculture, Ottawa, Ontario, Canada.

Tardif, P. and Caron, J. 1993. Unsaturated hydraulic conductivity of three peat substrates for nursery (in French). Internal report. Laval University, Quebec, Canada.

Topp, G.C. and Zebchuck, W. 1979. The determination of soil-water desorption curves for soil cores. *Can. J. Soil Sci.,* 59:19–26.

Topp, G.C., Davis, J.L., and Annan, A.P. 1980. Electromagnetic determination of soil water content: Measurement in coaxial transmission lines. *Water Resour. Res.,* 16:574–582.

Valat, B. 1989. Contribution to the study of organic materials used in horticulture (in French). Ph.D. thesis, University of Poitiers, France.

Valat, B., Jouany, C., and Rivière, L.M. 1991. Characterization of the wetting properties of air-dried peat and composts. *Soil Sci.,* 152:100–107.

Van Dijk, H. and Boekel, P. 1965. Effect of drying and freezing on certain physical properties of peat. *Neth. J. Agric. Sci.,* 13:248–260.

Van Lierop, W. 1983. Lime requirement determination of acid organic soils using buffer pH methods. *Can. J. Soil Sci.,* 63:411–423.

Verdonck, O. and Gabriëls, R. 1991. Substrates for horticultural crops, in *Proc. Symp. on Peat and Peatlands Diversification and Innovation.* Quebec, Canada, Aug. 6–10, 1989. Overend, R.P. and Jeglum J.K., Eds., Canadian Society for Peat and Peatlands, Dartmouth, Canada, 3–7.

Verhagen, J.B.G. 1997. Particle size distribution to qualify milled peat: a prediction of air content of ultimate mixtures, in *Peat in Horticulture. Proc. Int. Peat Conf.,* Amsterdam, The Netherlands, Nov. 2–7. Schmileski, G., Ed., 53–57.

Von Post, L. and Granlund, E. 1926. Peat Resources in South Sweden, I (in Swedish). *Sveriges Geologiska Undersökning,* C233 19:1–127.

Wallach, R.F.F., da Silva, F.F., and Chen, Y. 1992. Hydraulic characteristics of tuff (scocia) used as a container medium. *J. Amer. Soc. Hortic. Sci.,* 117:415–421.

Waller, P.L. and Harrison, A.M. 1991. Estimation of pore space and the calculation of air volume in horticultural substrates. *Acta Hortic.,* 294:29–39.

Warncke, P.D. and Krauskopf, D.M. 1983. Greenhouse growth media: Testing and nutrition guidelines. Ext. Bull. E1736, Michigan State University, East Lansing, Michigan.

Wever, G. 1995. Physical analysis of peat and peat-based growing media. *Acta Hortic.* 401:561–567.

White, J.W. 1965. The concept of container capacity and its application to soil moisture fertility regimes in the production of container grown crops. Ph.D. thesis, Penn. State University, University Park, Penn.

Woods, M.J. and Maher, M.J. 1971. Successive tomato cropping in peat substrates. *Acta Hortic.,* 18:80–84.

CHAPTER 5

Soil Acidity Determination Methods for Organic Soils and Peat Materials

Léon E. Parent and Catherine Tremblay

CONTENTS

ABSTRACT

Soil pH is one of the most important and frequently measured chemical indicators for assessing the quality of a soil for plant growth and microbial activity. The pH is determined using different suspension media, soil-to-solution ratios, and soil moisture conditions; therefore, conversion equations are presented in this chapter among methods for pristine and cultivated organic soil materials in this chapter. The 0.01 M CaCl$_2$ suspension was found to be the most reliable pH determination method against variations in moisture content and initial dewatering conditions of peat and moorsh materials. Compared with field-moist pristine peat materials, the air- or oven-drying procedure decreased pH values by 0.37 in water, 0.08 in 0.01 M CaCl$_2$, and 0.08 in 1 M KCl. The Shoemaker–McLean–Pratt buffer pH value was sensitive to the moorsh-forming process. Whereas peat oven-drying (105°C) increased buffer pH by 0.22–0.25 pH unit, moorsh oven-drying decreased buffer pH by 0.12 pH unit compared with field-moist conditions. Change in buffer pH upon drying should be further investigated to quantify the intensity of the moorsh-forming process in organic soils after drainage and reclamation.

I. INTRODUCTION

One of the most important and frequently measured chemical indicators for assessing the quality of a soil for plant growth and microbial activity is pH. Peat, made of more than 30% soil organic matter (SOM), is the constituting material of organic soils and is generally acidic. Peat materials form a complex of polycarboxylic acids that exchange protons with cationic species on an equivalent basis (Bloom and McBride, 1979). The origin of acidity in organic soil materials is related to the degree of decomposition and botanical makeup of the peat. Long-chain uronic acids rich in carboxyls predominate in fibric *Sphagnum* peat (Theander, 1954; Clymo, 1964). Humic and fulvic acids predominate in sapric peat (Naucke et al., 1990).

Lucas and Davis (1961) found the ideal pH (water suspension) for plant growth to be 5.0 in *Sphagnum* peat, and 5.5–5.8 in wood-sedge peat. Hoffmann (1964) specified a pH target of 4.3 in 1 M KCl for muck materials rich in clay and silt. Kuntze and Bartels (1984) recommended a target pH (0.1 M CaCl$_2$) of 4.5 for organic soils showing 30–60% SOM and rich in clay and silt; they also recommended a target pH of 4.0 for organic soils showing 30–60% SOM and rich in sand, as well as for organic soils containing more than 60% SOM. Optimum pH (1 M KCl) value was found to be 4.0 for onions and lettuce grown in organic soils (Van Lierop and MacKenzie, 1975; Van Lierop et al., 1980).

Many factors influence the value of soil pH, such as moisture conditions, soil-to-solution ratios, electrolyte concentrations, and the liquid junction potential (Huberty and Haas, 1940; Peech et al., 1953; Schofield and Taylor, 1955; Collins et al., 1970). The objective of this chapter is to document factors affecting pH measurements in organic soil materials, and to present conversion equations among determination methods.

II. FACTORS INFLUENCING SOIL PH VALUES

A. Soil-Solution Equilibration Time

Davis and Lawton (1947) equilibrated 50 field-moist moorsh samples (soil-to-solution ratio of 1:1 v/v) with distilled water for 1, 15, or 60 min. Differences were small but significant, with average pH values of 5.70, 5.69, and 5.67, respectively. Pessi (1962) found that the pH of eight field-moist pristine *Sphagnum* and forest sedge peat materials decreased by 0.07 pH unit between 30 min and 3 h after mixing, and stabilized after 7 h of equilibration. For air-dry mineral soils, a satisfactory pH stability was reached after the first 1–2 h of equilibration (Ryti, 1965). Van Lierop and MacKenzie (1977) did not report any significant effect of equilibration time (15 min vs. 60 min) on pH values of 10 field-moist or oven-dry moorsh samples. Van Lierop (1981a) selected a 30-min equilibration period for pH measurements of 30 field-moist moorsh samples.

B. Sampling Period

Van der Paauw (1962) found that the pH of mineral soils decreased in periods of low rainfall and increased in periods of high rainfall. Collins et al. (1970) found an average seasonal variation of 0.8 pH unit for water suspension of field-moist mineral soils, decreasing between June and September. In a *Sphagnum* peat limed at rates of 0 to 4 Mg ha^{-1}, Pessi (1962) found an average pH increase of 0.39 pH unit in water suspension and 0.09 pH unit in a 0.1 M CaCl$_2$ suspension, between June and October. In the long run, Hamilton and Bernier (1973) found that pH (0.01 M CaCl$_2$) of an acid organic soil (initial pH of 3.57 in a 0.01 M CaCl$_2$ suspension in control plots) tended to stabilize at 3.93, 4.48, 4.88, and 5.58 about 1.5 to 2 years after lime application at rates of 0, 6.7, 13.4, and 26.8 Mg ha^{-1}, respectively.

C. Soil-to-Solution Ratio

Peat bulk density is highly variable and poses a considerable challenge to soil testing (Van Lierop, 1981b), including pH determination methods (Van Lierop and MacKenzie, 1977). Davis and Lawton (1947) found a significant effect of three soil-to-solution ratios on water pH values of 15 field-moist moorsh samples ($P < 0.001$): the pH values averaged 5.36, 5.42, and 5.52 for the 1:0.5, 1:1, and 1:2.5 ratios, respectively. Ryti (1965) used a 1:2.5 soil-to-solution volumetric ratio in a pH study on air-dry pristine *Sphagnum-Carex* and wood-*Carex* peat samples. Van Lierop and MacKenzie (1977) used 1:2 to 1:4 soil-to-solution volumetric ratios for field-moist moorsh materials.

Van Lierop (1981a) obtained highly significant (R^2 = 0.998 to 0.999) relationships between 1:2 and 1:4 volumetric ratios on pH values in 0.01 M CaCl$_2$ and 1 M KCl suspension media between 3 and 7 for 30 field-moist moorsh samples. On average, differences were 0.024 and 0.110 pH unit for 0.01 M CaCl$_2$ and 1 M KCl, respectively. For the 0.01 M CaCl$_2$ suspension, the relationship was as follows:

$$Y = 0.995X + 0.049 \qquad (5.1)$$

For the 1 M KCl suspension, the relationship was as follows:

$$Y = 0.98X + 0.21 \qquad (5.2)$$

where Y is the 1:2 ratio and X is the 1:4 ratio.

Vaillancourt et al. (1999) compared 68 pH values of field-moist pristine peat materials determined according to volume (1:4 v/v) and weight (3:50 w/v) ratios (AOAC 1997). On average, the volume ratio (1:4 v/v) corresponded to a 1.4:50 (w/v) air-dry weight to solution ratio, which was nearly half the ratio recommended by AOAC (1997). Average pH differences between preparation methods (pH using 1.4:50 minus pH using 3:50 as air-dry weight to solution ratios) were 0.14 ± 0.07 for water, 0.04 ± 0.02 for 0.01 M CaCl$_2$, and 0.12 ± 0.02 for 1 M KCl. Thus, the 0.01 M CaCl$_2$ suspension showed the smallest pH difference, in keeping with Van Lierop (1981a). These results support the view that the 0.01 M CaCl$_2$ suspension would provide more stable pH values in peat and moorsh materials across a wider range of soil-to-solution ratios when compared with water or 1 M KCl suspension media.

D. Moisture Condition

Rost and Feiger (1923) found that pH values of 17 acid mineral soils dropped by 0.42 and 0.85 pH unit for air-dry and oven-dry samples, respectively, compared to field-moist samples. Comparatively, the pH of four alkaline mineral soils decreased by 0.54 and 0.32 pH units. The pH drop upon drying was irreversible even when the soil was moistened again and equilibrated for a long period of time. Soil pH should thus be taken under field-moist or natural conditions. Collins et al. (1970) observed a larger pH drop in water than in 0.01 M CaCl$_2$ suspensions due to air- or oven-drying 13 mineral soils.

Moisture conditions also influence the pH values of peat and moorsh materials (Davis and Lawton, 1947; Kaila et al., 1954). Kaila et al. (1954) emphasized the fact that the air-drying of peat samples can alter the properties of peat colloids, and the grinding necessary for the homogeneity of the material makes the conditions even more unnatural. Davis and Lawton (1947) compared the pH values of 15 field-moist and air-dry moorsh samples equilibrated for 1 or 15 min. Average pH values were 5.42 for field-moist samples and 5.34 for air-dry samples. Van Lierop and MacKenzie (1977) compared the pH values of 10 field-moist or oven-dry moorsh samples. They found that average pH values were lowered by 0.5, 0.2, and 0.2 pH unit in water, 0.01 M CaCl$_2$, and 1 M KCl, respectively. They attributed lower pH readings with dried samples to a significantly greater weight when soil solution ratios were measured volumetrically due to an increase in bulk density upon drying. Comparison between moisture conditions should thus be conducted on the same weight basis.

E. Suspension Media

Schofield and Taylor (1955) found that the lime potential of a mineral soil suspended in a 0.01 M CaCl$_2$ solution was a constant value computed as 1.14. Ryti (1965) presented survey data for air-dry soils where the average difference was 0.49 between pH values in water and 0.01 M CaCl$_2$ suspensions. A constant pH drop was observed only for loam and silt soils. Two air-dry pristine peat materials, a *Sphagnum-Carex* peat (pH 4.18) and a wood-*Carex* peat (pH 4.30), showed a similar pH drop of 0.45. For other soils (sand, clay, humus), the pH decrease was not constant: the higher the soil pH, the smaller was pH drop (Ryti, 1965). Davies (1971) showed, with equations from survey data of mineral soils, that the mean difference of 0.6 pH unit between pH values in water and in 0.01 M CaCl$_2$ decreased as soil pH increased.

In 20 field-moist moorsh materials, Van Lierop and MacKenzie (1977) found a pH drop of 0.55 between pH values in water and in 0.01 M CaCl$_2$, but the decrease was greater when soil pH in water was less than 4 compared with higher values (0.77 vs. 0.37 pH unit). In 30 field-moist moorsh materials, Van Lierop (1981a) found that pH in a water suspension was 0.44 pH unit higher than pH in 0.01 M CaCl$_2$, and 0.70 pH unit higher than pH in 1 M KCl. As mentioned previously, the higher the soil pH, the smaller the pH drop was. The conversion equations for field-moist moorsh materials between pH in water (1:4 v/v ratio) and pH in salt solutions (1:1 to 1:2 w/v ratios) were as follows (R^2 values of 0.98):

$$\text{Water pH} = 0.53 + 0.98 \text{ (pH in 0.01 } M \text{ CaCl}_2) \tag{5.3}$$

$$\text{Water pH} = 0.876 + 0.961 \text{ (pH in 1 } M \text{ KCl)} \tag{5.4}$$

III. EXPERIMENTAL SETUP

A. Interactions between Suspension Media and Moisture Conditions

Drainage and cultivation cause irreversible drying and pulverization, as well as chemical transformations in the peat (Pons, 1960). Peat forms hard aggregates or small particles upon drying, and the moorsh-forming process generates small particles (Volarovich et al., 1969; Meyerovsky and Hapkina, 1976). Irreversible drying during the moorsh-forming process probably involves the formation of hydrogen bonds, molecular condensation reactions, and metal complexes, thus affecting exchangeable hydrogen.

The authors examined the effect of the same weight of field-moist, air-dry or oven-dry (105°C) pristine peat (1.2 g in 20 ml, 71 samples) and moorsh (5 g in 20 mL, 48 samples) materials on pH values in water, 0.01 M CaCl$_2$, and 1 M KCl after 30 min of equilibration. The analysis of variance indicated significant interaction ($P< 0.001$) between moisture conditions and suspension media for both groupings. Statistics and contrasts are presented in Table 5.1.

Air- and oven-dry samples produced nonsignificant pH differences across suspension media in pristine peat materials and in the 1 M KCl for moorsh materials. Compared with field-moist pristine peat materials, air- or oven-dry samples decreased pH values on average by 0.37 in water, 0.08 in 0.01 M CaCl$_2$, and 0.08 in 1 M KCl. In moorsh materials, drying also reduced pH values compared with the field-moist condition, but air-dry materials produced intermediate results between field-moist and oven-dry samples in water and 0.01 M CaCl$_2$ suspension media. Drying affected water and 0.01 M CaCl$_2$ suspension pH values to a larger extent in pristine peat compared with moorsh materials (Table 5.1).

Table 5.1   Average pH Values of Peat and Moorsh Materials as Influenced by Preparation Method

Moisture Condition	Suspension (pH Unit)			Probability Level	
	Water	0.01 M CaCl$_2$	1 M KCl	Water vs. Salt	0.01 M CaCl$_2$ vs. 1 M KCl
Pristine Peat Materials (n = 71, s$_e$ = 0.218)					
Field-moist (FM)	4.73	3.69	3.48	<0.01	<0.01
Air-dry (AD)	4.34	3.62	3.41	<0.01	<0.01
Oven-dry (OD)	4.34	3.60	3.39	<0.01	<0.01
Probability Level					
FM vs. AD	<0.01	<0.01	<0.01	—	—
AD vs. OD	NS	NS	NS	—	—
Moorsh Materials (n = 48, s$_e$ = 0.080)					
Field-moist (FM)	6.21	5.86	5.66	<0.01	<0.01
Air-dry (AD)	6.13	5.81	5.59	<0.01	<0.01
Oven-dry (OD)	6.04	5.77	5.59	<0.01	<0.01
Probability Level					
FM vs. AD	<0.01	<0.01	<0.01	—	—
AD vs. OD	<0.01	<0.05	NS	—	—

Note: FM = field-moist, AD = air-dried, and OD = oven-dried.

B. Validation of Conversion Equations

Equations have been proposed to convert pH values of field-moist peat materials taken in distilled water, 0.01 M CaCl$_2$, and 1 M KCl suspensions (Van Lierop, 1981a; Vaillancourt et al., 1999). Little information is available for converting pH values among suspension media as a function of moisture conditions, using comparable soil-to-solution ratios. Relationships between pH values of pristine and moorsh materials are presented in Tables 5.2 and 5.3. The water suspension pH and the field-moist conditions of pristine peat materials produced the smallest coefficients of determination (r^2). These differences could be attributed to irreversible processes at microaggregate or colloidal scales occurring more intensively and heterogeneously in pristine peat than in moorsh materials. The authors' equation for air-dried pristine peat materials relating pH in 1 M KCl to pH in water gave a close fit with data from Kaila et al. (1954) (Figure 5.1), but overestimated observed values of water suspension pH by 0.06 pH unit in average. The fit was smaller using a larger data set from Kivekäs and Kivinen (1959) (Figure 5.2), and the predicted values of water pH were 0.05 pH unit lower than observed values; however, equations relating pH in 0.01 M CaCl$_2$ or 1 M KCl to pH in water underestimated water pH of field-moist samples by 0.38 to 0.63 pH unit. As a result, the conversion equations presented here should be validated before use.

Table 5.2   Conversion Equations between Soil pH Values in Three Suspension Media under Three Moisture Conditions

Suspension	Y Range	X Range	Equation ($y = aX + b$)	R^2
Pristine Peat Materials				
Water	3.79–6.02	3.25–5.73	$pH_{FM} = 0.864pH_{AD} + 1.053$	0.849
Water	3.79–6.02	3.19–5.90	$pH_{FM} = 0.850pH_{OD} + 1.087$	0.828
Water	3.25–5.73	3.19–5.90	$pH_{AD} = 0.967pH_{OD} + 0.113$	0.941
0.01 M CaCl$_2$	2.74–5.76	2.64–5.36	$pH_{FM} = 1.024pH_{AD} + 0.034$	0.962
0.01 M CaCl$_2$	2.74–5.76	2.69–5.64	$pH_{FM} = 1.041pH_{OD} - 0.072$	0.942
0.01 M CaCl$_2$	2.64–5.36	2.69–5.64	$pH_{AD} = 0.994pH_{OD} + 0.018$	0.962
1 M KCl	2.44–5.65	2.35–5.39	$pH_{FM} = 1.009pH_{AD} + 0.122$	0.978
1 M KCl	2.44–5.65	2.36–5.61	$pH_{FM} = 0.999pH_{OD} + 0.108$	0.981
1 M KCl	2.35–5.39	2.36–5.61	$pH_{AD} = 0.978pH_{OD} + 0.026$	0.979
Moorsh Materials				
Water	4.93–7.25	4.97–7.10	$pH_{FM} = 1.022pH_{AD} - 0.060$	0.975
Water	4.93–7.25	4.92–7.04	$pH_{FM} = 1.048pH_{OD} - 0.127$	0.935
Water	4.97–7.10	4.92–7.04	$pH_{AD} = 1.028pH_{OD} + 0.080$	0.964
0.01 M CaCl$_2$	4.85–6.88	4.79–6.81	$pH_{FM} = 1.002pH_{AD} - 0.037$	0.985
0.01 M CaCl$_2$	4.85–6.88	4.80–6.71	$pH_{FM} = 1.016pH_{OD} - 0.041$	0.957
0.01 M CaCl$_2$	4.79–6.81	4.80–6.71	$pH_{AD} = 1.019pH_{OD} - 0.070$	0.981
1 M KCl	4.61–6.72	4.52–6.63	$pH_{FM} = 0.984pH_{AD} + 0.164$	0.980
1 M KCl	4.61–6.72	4.55–6.55	$pH_{FM} = 1.030pH_{OD} + 0.091$	0.960
1 M KCl	4.52–6.63	4.55–6.55	$pH_{AD} = 1.050pH_{OD} - 0.276$	0.986

Note: FM = field-moist; AD = air-dried; and OD = oven-dried.

Table 5.3   Conversion Equations between Soil pH Values among Suspension Media for Field-Moist, Air-Dry and Oven-Dry Peat and Moorsh Samples

Moisture Condition	Y Range	X Range	Equation ($y = aX + b$)	R^2
Pristine Peat Materials				
Field-moist	3.79–6.02	2.74–5.16	$pH_W = 0.852pH_{CaCl_2} + 1.670$	0.908
Field-moist	3.79–6.02	2.44–5.11	$pH_W = 0.776pH_{KCl} + 2.105$	0.896
Field-moist	2.74–5.16	2.44–5.11	$pH_{CaCl_2} = 0.912pH_{KCl} + 0.509$	0.989
Air dry	3.41–6.39	2.64–5.91	$pH_W = 0.851pH_{CaCl_2} + 1.276$	0.964
Air dry	3.41–6.39	2.35–5.90	$pH_W = 0.765pH_{KCl} + 1.756$	0.952
Air dry	2.64–5.91	2.35–5.90	$pH_{CaCl_2} = 0.900pH_{KCl} + 0.558$	0.993
Oven-dry	3.43–6.02	2.74–5.64	$pH_W = 0.870pH_{CaCl_2} + 1.247$	0.922
Oven-dry	3.43–6.02	2.39–5.61	$pH_W = 0.765pH_{KCl} + 1.808$	0.912
Oven-dry	2.74–5.64	2.39–5.61	$pH_{CaCl_2} = 0.882pH_{KCl} + 0.636$	0.996
Moorsh Materials				
Field-moist	4.93–7.25	4.68–6.88	$pH_W = 1.052pH_{CaCl_2} + 0.047$	0.952
Field-moist	4.93–7.25	4.45–6.72	$pH_W = 1.015pH_{KCl} + 0.458$	0.928
Field-moist	4.68–6.88	4.45–6.72	$pH_{CaCl_2} = 0.972pH_{KCl} + 0.355$	0.988
Air dry	4.97–7.10	4.78–6.81	$pH_W = 1.041pH_{CaCl_2} + 0.086$	0.955
Air dry	4.97–7.10	4.52–6.63	$pH_W = 0.996XpH_{KCl} + 0.562$	0.936
Air dry	4.78–6.81	4.52–6.63	$pH_{CaCl_2} = 0.964pH_{KCl} + 0.415$	0.995
Oven-dry	4.92–7.04	4.71–6.71	$pH_W = 1.033pH_{CaCl_2} + 0.086$	0.985
Oven-dry	4.92–7.04	4.55–6.55	$pH_W = 1.019pH_{KCl} + 0.348$	0.974
Oven-dry	4.71–6.71	4.55–6.55	$pH_{CaCl_2} = 0.990pH_{KCl} + 0.237$	0.995

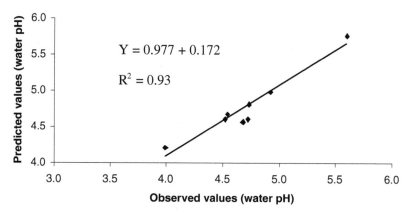

Figure 5.1 Relationship between predicted and observed water pH using the equation relating pH in 1 M KCl to water pH for air-dried pristine peat materials.

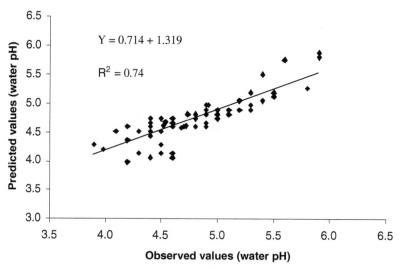

Figure 5.2 Relationship between predicted and observed water pH using the equation relating pH in 1 M KCl to water pH for air-dried pristine peat materials.

Table 5.4   Conversion Equations between SMP Soil Buffer pH Values of Peat (1.2 g of Peat in 20 mL of Suspension Medium) and Moorsh (5 g of Moorsh in 20 mL of Suspension Medium) Materials under Three Moisture Conditions

Y Range	X Range	Equation (y = aX + b)	R^2
Pristine Peat Materials			
4.24–6.84	4.29–6.80	$pH_{FM} = 0.951pH_{AD} + 0.241$	0.939
4.24–6.84	4.39–6.89	$pH_{FM} = 0.867pH_{OD} + 0.531$	0.800
4.29–6.80	4.39–6.89	$pH_{AD} = 0.928pH_{OD} + 0.217$	0.881
Moorsh Materials			
5.08–6.63	5.03–6.57	$pH_{FM} = 1.005pH_{AD} + 0.053$	0.991
5.08–6.63	5.02–6.51	$pH_{FM} = 1.055pH_{OD} - 0.214$	0.979
5.03–6.57	5.02–6.51	$pH_{AD} = 1.050pH_{OD} - 0.265$	0.988

Note: FM = field-moist; AD = air-dried; OD = oven-dried.

C. pH Indicator of the Moorsh-Forming Process

Change in exchangeable acidity can be assessed readily by routine buffer pH analysis (Shoemaker et al., 1961; Adams and Evans, 1962; Hajek et al., 1972). Thus, the alteration of peat and moorsh colloids and microaggregates upon drying could be documented by buffer pH determinations. Relationships between buffer pH values among moisture conditions across suspension media for pristine peat and moorsh materials are presented in Table 5.4. Curve fitting was closer with moorsh than pristine peat materials across moisture conditions, indicating more variable and intensive alteration in pristine peat upon drying.

In pristine peat materials, SMP buffer pH increased by 0.22–0.25 pH unit with oven-drying, whatever the initial suspension media (Table 5.5). Thus, exchangeable acidity decreased in pristine peat upon irreversible drying, presumably due to molecular condensation and formation of compact aggregates leading to increased hydrophobicity.

In contrast, buffer pH decreased by 0.12 pH unit in moorsh materials upon oven-drying, indicating enhanced exchangeable acidity in dried moorsh compared with moist moorsh. An appreciable fraction of humic molecules is not in contact with the solution in the aggregated structure of wet soil materials (Raveh and Avnimelech, 1978). When the soil is dried, the micro-aggregated structure of humic substances due to hydrogen bonds in the presence of water is broken, and the stability of organic matter decreases leading to the dispersion of the organic matrix (Raveh and Avnimelech, 1978). Presumably, high-molecular-weight organic molecules and microaggregates collapsed upon drying moorsh materials, thus exposing protons.

Buffer pH differences between field-moist and either air- or oven-dry samples could be a useful indicator of peat transformation into moorsh, as well as the degree of irreversible drying. The more negative the difference, the higher the degree of water repellency upon drying, and the less advanced the moorsh-forming process. The more positive the difference, the more ripened the moorsh would be.

Table 5.5   Average Buffer pH Values of Peat and Moorsh Materials as Influenced by the Preparation Method

Moisture Condition	Suspension (pH Units)			Probability Level	
	Water	0.01 M CaCl$_2$	1 M KCl	Water vs. Salt	0.01 M CaCl$_2$ vs. 1 M KCl
Pristine Peat Materials (n = 71, s$_e$ = 0.158)					
Field-moist (FM)	5.43	5.44	5.45	NS	NS
Air-dry (AD)	5.51	5.51	5.53	NS	NS
Oven-dry (OD)	5.68	5.64	5.67	NS	NS
Probability Level					
FM vs. AD	*P < 0.01*	*P < 0.01*	*P < 0.01*	—	—
AD vs. OD	*P < 0.01*	*P < 0.01*	*P < 0.01*	—	—
Moorsh Materials (n = 48, s$_e$ = 0.041)					
Field-moist (FM)	6.20	6.20	6.26	*< 0.01*	*< 0.01*
Air-dry (AD)	6.13	6.11	6.11	*< 0.01*	NS
Oven-dry (OD)	6.08	6.08	6.11	*< 0.05*	*< 0.01*
Probability Level					
FM vs. AD	*P < 0.01*	*P < 0.01*	*P < 0.01*	—	—
AD vs. OD	*P < 0.01*	*P < 0.01*	NS	—	—

Note: FM = field-moist; AD = air-dried; OD = oven-dried; NS = nonsignificant.

Source: From Shoemaker, H.E., McLean, E.O., and Pratt, P.F. 1961. *Soil Sci. Soc. Am. Proc.*, 25:274–277. With permission.

IV. CONCLUSION

The pH is an important driving variable for microbial and chemical processes in soil systems. Soil pH measurements have been conducted using different suspension composition, soil-to-solution ratios, and soil moisture conditions. Conversion equations should be examined carefully before interpreting results from different laboratories. The peat-forming process can be distinguished from the moorsh-forming process by the difference in SMP buffer pH between wet and oven-dry samples.

ACKNOWLEDGMENTS

This project was supported by the Natural Sciences and Engineering Research Council of Canada (OG #2254).

REFERENCES

Adams, F. and Evans, C.E. 1962. A rapid method for measuring lime requirement of red-yellow podzolic soils. *Soil Sci. Soc. Am. Proc.,* 26:355–357.

AOAC. 1997. AOAC official method 973.04: pH of peat, in *Official Methods of Analysis,* 16th ed. (1990), vol. 1, Suppl. March 1996. Cunnif, P., Ed., Association of Official Analytical Chemists International., Gaithersburg, Maryland.

Bloom, P.R. and McBride, M.B. 1979. Metal ion binding and exchange with hydrogen ions in acid-washed peat. *Soil Sci. Soc. Am. J.* 43:687–692.

Clymo, R.S. 1964. The origin of acidity in *Sphagnum* bogs. *Bryologist,* 67:427.

Collins, J.B., Whiteside, E.P., and Cress, C.E. 1970. Seasonal variability of pH and lime requirements in several southern Michigan soils when measured in different ways. *Soil Sci. Soc. Am. Proc.,* 34:56–61.

Davies, B.E. 1971. A statistical comparison of pH values of some English soils after measurement in both water and 0.01 *M* calcium chloride. *Soil Sci. Soc. Am. Proc.,* 35:551–552.

Davis, J.F. and Lawton, K. 1947. A comparison of the glass electrode and indicator methods for determining the pH of organic soils and effect of time, soil-water ratio, and air-drying on glass electrode results. *J. Amer. Soc. Agron.,* 39:719–723.

Hajek, B.F., Adams, F., and Cope, J.T., Jr. 1972. Rapid determination of exchangeable bases, acidity, and base saturation for soil characterization. *Soil Sci. Soc. Am. Proc.,* 36:436–438.

Hamilton, H.A. and Bernier, R. 1973. Effects of lime on some chemical characteristics, nutrient availability, and crop response of a newly broken organic soil. *Can. J. Soil Sci.,* 53:1–8.

Hoffmann, W. 1964. Influence of pH, phosphate fixation and soluble Al and Fe in organic soils on plant growth (in German). *Zeitschrift für Pflanzenernährung, Düngung und Bodenkunde,* 107:223–241.

Huberty, M.R. and Haas, A.R. 1940. The pH of soil as affected by soil moisture and other factors. *Soil Sci.,* 49:455–478.

Kaila, A., Soini, S., and Kivinen, E. 1954. Influence of lime and fertilizers upon the mineralization of peat nitrogen in incubation experiments. *J. Sci. Agric. Soc. Finland,* 26:79–95.

Kivekäs, J. and Kivinen, E. 1959. Observations on the mobilization of peat nitrogen in incubation experiments. *J. Sci. Agric. Soc. Finland,* 31:268–281.

Kuntze, H. and Bartels, R. 1984. Liming of cultivated organic soils (in German), in *Bewirtschaftung und Düngung von Moorböden.* Kuntze, H., Ed., Bremen Soil Institute, Bremen, Germany, 27–33.

Lucas, R.E. and Davis, J.F. 1961. Relationships between pH values in organic soils and availabilities of 12 plant nutrients. *Soil Sci.,* 92:177–182.

Meyerovsky, A.S. and Hapkina, Z.A. 1976. Transformation of Byelorussian peat soils as influenced by reclamation and agricultural use. *Proc. 5th Int. Peat Congr.,* 1:248–255.

Naucke, W., et al. 1990. Mire chemistry, in *Mires: Process, Exploitation and Conservation.* Heathwaite, A.L. and Göttlich, Kh., Eds., John Wiley & Sons, New York, 263–309.

Peech, M., Olsen, R.A., and Bolt, G.H. 1953. The significance of potentiometric measurements involving liquid junction in clay and soil suspension. *Soil Sci. Soc. Am. Proc.,* 17:214–218.

Pessi, Y. 1962. The pH-reaction of the peat in long-term soil improvement and fertilizing trials at the Leteensuo experimental station. *J. Sci. Agric. Soc. Finland,* 34:44–54.

Pons, L.J. 1960. Soil genesis and classification of reclaimed peat soils in connection with initial soil formation. *Proc. 7th Int. Congr. Soil Sci.,* 205–211.

Raveh, A. and Avnimelech, Y. 1978. The effect of drying on the colloidal properties and stability of humic compounds. *Plant Soil,* 50:545–552.

Rost, C. and Feiger, E.A. 1923. Effect of drying and storage upon the hydrogen-ion concentration of soil samples. *Soil Sci.,* 16:121–126.

Ryti, R. 1965. On the determination of soil pH. *J. Sci. Agric. Soc. Finland,* 37:51–60.

Schofield, R.K. and Taylor, A.W. 1955. The measurement of soil pH. *Soil Sci. Soc. Am. Proc.,* 19:164–167.

Shoemaker, H.E., McLean, E.O., and Pratt, P.F. 1961. Buffer methods for determining lime requirement of soils with appreciable amounts of extractable aluminium. *Soil Sci. Soc. Am. Proc.,* 25:274–277.

Theander, O. 1954. Studies on *Sphagnum* peat. 3. A quantitative study of the carbohydrate constituents of *Sphagnum* mosses and *Sphagnum* peat. *Acta Chem. Scand.,* 8: 989.

Vaillancourt, N. et al. 1999. Sorption of ammonia and release of humic substances as related to selected peat properties. *Can. J. Soil Sci.,* 79:311–315.

Van der Paauw, F. 1962. Periodic fluctuations of soil fertility, crop yields and of responses to fertilization effected by alternating periods of low and high rainfall. *Plant Soil,* 17:154–182.

Van Lierop, W. 1981a. Conversion of organic soil pH values in water, 0.01 *M* CaCl₂ or 1 N KCl. *Can. J. Soil Sci.,* 61:577–579.

Van Lierop, W. 1981b. Laboratory determination of field bulk density for improving fertilizer recommendations of organic soils. *Can. J. Soil Sci.,* 61:475–482.

Van Lierop, W., Martel, Y.A., and Cescas, M.P. 1980. Optimal soil pH and sufficiency concentrations of N, P and K for maximum alfalfa and onion yields on acid organic soils. *Can. J. Soil Sci.,* 60:107–117.

Van Lierop, W. and MacKenzie, A.F. 1977. Soil pH measurement and its application to organic soils. *Can. J. Soil Sci.,* 57:55–64.

Van Lierop, W. and MacKenzie, A.F. 1975. Effects of calcium carbonate and sulphate on the growth of lettuce and radish in some organic soils of southwestern Quebec. *Can. J. Soil Sci.,* 55:205–212.

Volarovich, M.P., Lishtvan, I.I., and Terent'ev, A.A. 1969. Electron-microscopic data on the highly dispersed fraction of peat. *Kolloidnyi Zhurnal,* 31:148–151.

CHAPTER 6

Nitrogen and Phosphorus Balance Indicators in Organic Soils

Léon E. Parent and Lotfi Khiari

CONTENTS

ABSTRACT

Nutrient losses from agroecosystems depend on the amount of water discharge, soil type, and management practices. This chapter presents N and P indicators of organic soil quality as related to water quality and crop fertilization. Organic soils contain 5 to 27 Mg organic N ha^{-1} in the arable layer, which could release 800 to 1500 kg NO$_3$-N ha^{-1} a^{-1}, depending primarily on C/N ratio and pH. Because 12 to 245 kg NO$_3$-N ha^{-1} a^{-1} could be discharged, and because crop removal cannot account for residual N, most of the N must be denitrified. Organic soils contain in average 700 to 1100 mg P kg^{-1}, of which 67 to 78% is reported to be organic P, inorganic P being more or less chemically sorbed. One to 88 kg P ha^{-1} would be discharged annually, indicating high potential risk for eutrophication. Due to the predominance of organic forms, N and P microbial turnovers should be diagnosed, possibly using N and P ratios or multiratios (e.g., C/N, N/P, C/N/P, C/N/P/S). An inorganic P sorption or saturation index, as well as N and P turnover attributes as multiratios, should be developed for planning N and P fertilization programs.

I. INTRODUCTION

Soil quality indicators can assist managers of agroecosystems in making profitable and environmentally sound decisions. Groffman et al. (1996) proposed the calibration of a suite of microbial variables that are useful as indices of important wetland nutrient cycling and water quality functions: microbial biomass indicating the capacity of an ecosystem to support nutrient cycling and biodegradation; denitrification enzyme activity as an index of soil denitrification capacity; N mineralization rate as an important component of soil fertility and an indicator of N availability; and soil respiration as an index of overall biological activity. They found that the mean water table level and soil organic matter content, which comprises the substrate for most microbial processes represented by soil total C and N, were the strongest predictors of C and N cycle variables in 12 New York wetlands.

Chemical indicators are assessed by soil and water analyses. Soil testing is indicative of soil fertility status and of the environmental impact of soil management (Khiari et al., 2000). Soil quality is related to water quality, because a determinant soil function is to store and transmit water. Nutrient concentrations in runoff water from a German organic soil showed 10-year average values of 9.75 mg Ca L^{-1}, 0.91 mg P L^{-1}, 9.10 mg K L^{-1} and 8.35 mg N L^{-1} (Kuntze and Eggelsmann, 1975). Coefficients of variations were 15 and 17% for K and Ca, respectively, compared with 48 and 67% for P and N, thus indicating greater influence of varying climatic conditions on P and N losses compared with K and Ca.

Drained and cultivated organic soils in the Florida Everglades discharge large amounts of plant nutrients into drainage canals (Volk and Sartain, 1976), and contribute significantly to lake eutrophication by N and P (Hortenstine and Forbes, 1972). The P losses from organic soils also contribute to eutrophication of Lake Ontario (Nicholls and MacCrimmon, 1974; Miller, 1979; Longabucco and Rafferty, 1989).

The aim of this chapter is to document soil N and P indicators for improving the management of cultivated organic soils in relation to water quality.

II. ENVIRONMENTAL RISK OF CULTIVATING ORGANIC SOILS

Nutrient loss depends on the amount of water discharge, soil type, and fertilization practices. Across 5 months in a dry season, Nicholls and MacCrimmon (1974) found small annual losses of 4.1 kg N ha^{-1} and 1.6 kg P ha^{-1} in an Ontario organic soil cultivated for 12 years. A New York organic soil released 0.6 to 30.7 kg P ha^{-1}, 39.2 to 87.5 kg NO_3-N ha^{-1}, and less than 1 to 1.9 kg NH_4-N ha^{-1} annually (Duxbury and Peverly, 1978). Miller (1979) found annual losses of 37 to 245 kg N ha^{-1}, 2 to 37 kg P ha^{-1}, and 288 kg K ha^{-1} in overfertilized Ontario organic soils cultivated for 30 years. These authors concurred that crops grown in organic soils received probably excessive amounts of N, P, and K fertilizers.

As an indication of P accumulation and N requirement, fertilizer trials conducted in old, cultivated organic soils in New York (Minotti and Stone, 1988) showed no adverse consequence to onion crops of omitting P fertilizers for 3 to 4 years; however, a significant N requirement existed during 4 years out of 8 for early planted onions even though substantial amounts can be released later in the season.

The P losses by leaching from organic soils are several orders of magnitude larger than for mineral soils, and depend on the amounts of mineralization and fertilization that have occurred as well as on the ability of the soil to sorb P (Cogger and Duxbury, 1984). The P leaching reportedly varied from 0.6 to 36 kg P ha^{-1} in organic soils (Cogger and Duxbury, 1984; Scheffer and Kuntze, 1989; Porter and Sanchez, 1992). Such high P loss was attributed to small amounts of Ca, Al, and Fe compounds reacting with inorganic P in organic soils (Miller, 1979; Scheffer and Kuntze, 1989; Porter and Sanchez, 1992). Miller (1979) suggested considering P sorption capacity in organic soil use and management.

III. FERTILIZER TRIALS CONDUCTED IN QUEBEC

In view of reducing N and P inputs in organic soils, N and P fertilization trials have been conducted during 3 years (1994–1996) in organic soils of southwestern Quebec on high-value crops such as onions (*Alium cepa* L.), celery (*Apium graveolens* L.), and carrots (*Daucus carota* L.) (Asselin, 1997). The four N levels were: control (no N), half the rate recommended by the provincial authority (CPVQ, 1996), the recommended rate, and 1.5 times the recommended rate. Two methods of application were used: a broadcast application before sowing and a row application later in the season. Three P levels, according to a soil test (Mehlich, 1984), were: control (no P), half the recommended rate, and the recommended rate (CPVQ, 1996). The P was applied broadcast before sowing. The eight treatments are presented in Table 6.1. The treatments were arranged in a randomized block design with four replicates.

Table 6.1   The N and P Fertilization Treatments in the Quebec Experiment

N Treatments (kg N ha⁻¹)		P Treatments (kg P ha⁻¹)		
Before Sowing	After Emergence	1994	1995	1996
Onion (*Allium cepa* L.)				
0	0	0	0	0
0	0	15	40	44
43	0	15	40	44
43	43	15	40	44
87	0	15	40	44
87	43	15	40	44
87	43	7.5	20	22
87	43	0	0	0
Celery (*Apium graveolens* L.)				
0	0	0	0	0
0	0	27	35	15
37	0	27	35	15
37	37	27	35	15
73	0	27	35	15
73	37	27	35	15
73	37	13.5	17.5	7.5
73	37	0	0	0
Carrot (*Daucus carota* L.)				
0	0	0	0	0
0	0	20	20	40
17	0	20	20	40
17	17	20	20	40
33	0	20	20	40
33	17	20	20	40
33	17	10	10	20
33	17	0	0	0

Source: From Asselin, M. 1997. Computer model for rational use of fertilizers and the correction of nutrient imbalances in vegetable crops on organic soils (in French). Canada-Quebec Agreement Sustain. Environ. Agric. Rep. 13–67130811–046, Quebec. With permission.

The dataset comprised soil classification, early plant growth as fresh weight to test starter fertilizer effects, biomass production, and marketable yield as fresh weight at harvest, soil P (Mehlich, 1984), as well as pH and nitrate ($N\text{-}NO_3$) and phosphate ($P\text{-}PO_4$) concentrations in the saturated paste (Warncke, 1990). Surface (0–20 cm) samples were collected in the designated fields during the fall preceding the trial. Soil samples were sent to a certified laboratory, where they were dried at 105°C, ground to <2 mm, and 3-mL scooped before extraction by the Mehlich-III solution. Soil samples were also taken 4 to 5 times during the growing season at six depths (0–20, 20–40, 40–60, 60–80, 80–100, and 100–120 cm), and analyzed for soluble nitrate and phosphate after saturating the soil with distilled water and extracting

Table 6.2   Soil Classification and Median Values in the 0–40 cm Soil Layers for Soil Paste pH across Season and Soil Bulk Density

Year	Crop	Soil Class[a]	Moisture Class[b]	Moorsh Class[b]	pH (H_2O)	Bulk Density (g cm^{-3})	Mehlich-III P (kg ha^{-1})
1994	Onion	Typic Mesisol	C	MtIIIcb	5.83	0.185	310
	Celery	Limnic Mesisol	BC	MtIIbc	5.97	0.162	125
	Carrot	Limnic Mesisol	C	MtIIIbc	5.11	0.186	215
1995	Onion	Limnic Fibrisol	BC	MtIIac	5.59	0.203	120
	Celery	Fibric Mesisol	C	MtIIIba	6.55	0.325	70
	Carrot	Mesic Fibrisol	C	MtIIIba	6.34	0.253	220
1996	Onion	Humic Mesisol	C	MtIIIcb	5.83	0.313	67
	Celery	Terric Humisol	C	MtIIIc4	5.02	0.468	380
	Carrot	Limnic Mesisol	C	MtIIIbc	5.29	0.232	75

Note: C = dry condition; BC = moderately dry conditions.
[a] According to Agriculture Canada. 1992.
[b] According to Okruszko and Ilnicki (Chap. 1, this volume).

under vacuum using the saturated paste method (Warncke, 1990). Soil pH was determined by inserting the electrode directly into the paste. Bulk density was determined by the cylinder method. The soils were classified according to Agriculture Canada (1992) and Okruszko and Ilnicki (Chap. 1, this volume) (Table 6.2).

The authors conducted ANOVA analyses (SAS Institute, 1989) on fresh biomass and yield data. Nitrate and phosphate data were transformed into kg ha^{-1} for each stratum after accounting for moisture content and bulk density. These N and P values were combined across depths as N and P accumulated in the 0–40, 40–80, and 80–120 cm layers, respectively. Due to high variability, we selected median N and P values from 360 to 1440 determinations at each N or P level across crops, years, and replicates in space and time. Median values were expressed as relative enrichment due to fertilization compared with control, in order to obtain indications of the relative effect of fertilization on N and P accumulation in the soil profile.

The crops yielded differently over the 3 years of experimentation (Table 6.3). In particular, carrots yielded highest in 1994 and 1996, and onions in 1995. Celery yields were highest in 1996. Onions and carrots did not respond significantly to N or P fertilization. Only celery responded to N in 2 out of 3 years.

Hamilton and Bernier (1975) obtained similar results in the 1965–1968 period; their experimental site had been cleared 4 years previous to the commencement of the experiment, and cropped twice to potato (*Solanum tuberosum* L.) and once to celery with 22.4 kg N ha^{-1} and 48 kg P ha^{-1}, then left uncropped and unfertilized in the year preceding the experiment. Their median yields were 33 Mg ha^{-1} for onions, 31 Mg ha^{-1} for celery, and 45 Mg ha^{-1} for carrots. Onions did not respond to either N, P, or K. Carrots and celery responded to K only. Obviously, the non-response to N in Quebec organic soils compared with significant response to N in those from western New York south of Lake Ontario (Minotti and Stone, 1988) is attributable to yield difference. Onion yields were 38–55 Mg bulbs ha^{-1} in our trial (Table 6.3) and 58–71 Mg bulbs ha^{-1} in the Minotti and Stone (1988) experiments.

Table 6.3 Fresh Yield of Biomass and Marketable Products as Related to Fertilization in Organic Soils of Southwestern Quebec

Year	Early Biomass (Starter Effect)				Late Biomass at Harvest				Marketable Product			
	Control[a] (kg ha⁻¹)	Fertilized (kg ha⁻¹)	F value	C.V. (%)	Control[a] (kg ha⁻¹)	Fertilized (kg ha⁻¹)	F value	C.V. (%)	Control[a] (kg ha⁻¹)	Fertilized (kg ha⁻¹)	F value	C.V. (%)
					Onion (*Allium cepa* L.)							
1994	1232	1303	0.66NS	20.3	36,109	43,676	1.43NS	13.5	—	—	—	—
1995	5099	3797	1.11NS	27.6	79,817	75,503	1.31NS	10.0	48,607	54,573	0.91NS	13.6
1996	3282	2867	0.98NS	24.2	66,025	75,154	1.14NS	16.4	41,320	37,830	1.60NS	20.8
Average	3204	2656	—	—	60,650	64,628	—	—	44,964	46,201	—	—
					Celery (*Apium graveolens* L.)							
1994	934	1491	6.39**	12.7	93,659	94,334	1.75NS	10.8	—	—	—	—
1995	2519	3154	0.80NS	23.3	89,035	91,530	1.17NS	14.3	48,733	50,580	1.18NS	16.4
1996	7254	9122	3.28**	14.2	113,666	136,360	2.51*	10.7	62,180	87,701	2.77*	15.3
Average	3569	4589	—	—	98,786	107,408	—	—	55,457	69,141	—	—
					Carrot (*Daucus carota* L.)							
1994	1952	1893	0.26NS	23.8	123,077	119,391	0.93NS	11.3	—	—	—	—
1995	2235	2230	0.60NS	26.1	63,022	71,608	0.65NS	17.4	39,263	46,980	0.28NS	25.2
1996	9966	8703	0.08NS	32.4	148,774	143,773	0.24NS	18.4	77,331	71,614	0.39NS	21.2
Average	4718	4276	—	—	111,624	111,591	—	—	58,297	57,797	—	—

Note: NS, *, **: nonsignificant and significant at the 0.05 and 0.01 levels, respectively.
[a] No N and P added.

Source: From Asselin, M. 1997. Computer model for rational use of fertilizers and the correction of nutrient imbalances in vegetable crops on organic soils (in French). Canada-Quebec Agreement Sustain. Environ. Agric. Rep. 13-67130811-046, Quebec. With permission.

Table 6.4   Median Percentage Values of Nitrate and Phosphate Distribution in the Soil Profile and of N and P Accumulation at Recommended Rates over Control Receiving No N and P Fertilizers

Depth (cm)	N Distribution in the Soil Profile (%)			P Distribution in the Soil Profile (%)		
	Onion	Celery	Carrot	Onion	Celery	Carrot
0–40	61	60	60	49	57	45
40–80	26	25	30	32	30	35
80–120	13	15	10	19	13	20
	Accumulated N over Control (%)			Accumulated P over Control (%)		
	Onion	Celery	Carrot	Onion	Celery	Carrot
0–40	29	34	0	13	29	8
40–80	13	1	1	1	12	6
80–120	44	20	−2	−1	39	0
0–120	23	9	0	3	19	6

The nonresponse to P across the previously described trials reflects sufficient release of available P in organic soils even at high yield level.

As a result, the recommended N and P rates in Quebec organic soils appeared excessive both economically and agronomically, and were environmentally at risk: 60% of the soil nitrate and 50% of the soil soluble phosphate were accumulated in the 0–40 cm top layer, thus leaving large proportions beyond the rooting zone (Table 6.4). Onions and celery were the most environmentally at risk for N and P nonpoint pollution. The carrot crop produced the smallest N accumulation in the soil (Table 6.4) due to smaller N application rate and, possibly, to greater rhizosphere capacity for denitrification. Excessive P fertilization was most problematic in the celery crop, due to a combination of low phosphate retention capacity in organic soils and water supplied through sprinkler irrigation. The N budget indicated considerable wastage of N fertilizers across crops (Table 6.5), because significant response to N was obtained only in celery at the lowest N rate (37 kg N ha^{-1}). The P budget showed that carrots had greater P removal capacity compared with onions and celery. The P turnover of specific soil–plant systems at each site, as well as irrigation facilitating the P leaching, are probably involved in the P distribution across the soil profile.

Consequently, recommending N and P according to the present guidelines based on concepts such as build-up and maintenance, nutrient uptake by the crop, or soil

Table 6.5   Removal of N and P by Three Vegetable Crops Grown in Organic Soils at Recommended N and P Rates

Year	Crop	N Rate Fertilizer (kg N ha^{-1})	N Uptake Foliage (kg N ha^{-1})	N Uptake Harvest (kg N ha^{-1})	P Rate Fertilizer (kg P ha^{-1})	P Uptake Foliage (kg P ha^{-1})	P Uptake Harvest (kg P ha^{-1})
1995	Celery	110	84	75	35	14	29
	Carrot	50	119	82	20	23	34
1996	Onion	130	84	75	44	14	29
	Carrot	50	171	120	40	52	64

Source: From Asselin, M. 1997. Computer model for rational use of fertilizers and the correction of nutrient imbalances in vegetable crops on organic soils (in French). Canada-Quebec Agreement Sustain. Environ. Agric. Rep. 13–67130811–046, Quebec. With permission.

testing, with methods developed for mineral soils, may lead to excessive water pollution in organic soil farming. Because organic P makes up 67% of total P in Quebec organic soils (Parent et al., 1992), its turnover should be given more attention. Organic soil subsidence by the irreversible biological decomposition of the organic matter in drained and aerated organic soil layers releases significant seasonal quantities of organic N and P not accounted for by soil testing.

Indeed, the very poor water quality of the Norton Creek draining the investigated soil area has been attributed primarily to its high P concentration (i.e., 0.28 to 0.60 mg total P L^{-1}) with median concentration exceeding 14 times the environmental standard of 0.03 mg L^{-1} (Simoneau, 1996). Because half of total N was in the form of nitrate and nitrite, and because dissolved P concentration was 9 times that of particulate P, runoff and leaching of fertilizers applied in large amounts to vegetable crops were believed to contribute appreciably to the N and P pollution of the Norton Creek. This requires a substantial reduction in N and P fertilizer recommendations, and to export N and P out of the soil system by the harvested portion of the crop. Soil–plant diagnoses should be improved in organic soils based on the N and P soil cycles in order to reduce N and P discharge to surface water while maintaining these soils as highly productive.

IV. SIMPLIFIED N AND P CYCLES

The requirements for building nutrient soil cycle models are (Frissel, 1977):

1. Knowledge of the elements under examination, such as water solubility (N, P), volatility (N), and degree of chemical reactivity (P)
2. Nature and size of compartments, and balance among them
3. Pathways and rates of transfer
4. Reference time period for processes
5. Definition of the area and boundaries of the system

Nutrients may be lost by leaching or volatilization, or taken up by the crop (Figure 6.1). A simplified internal soil cycle contains three compartments releasing nutrients into the soil solution.The amounts of nutrients removed are controlled by soil moisture, temperature, organic matter content, acidity, aeration, depth to impermeable layer, patterns and seasonality of rainfall, microbial interactions, and cultural operations (Frissel, 1977). Human influence is exerted primarily through soil management, as well as selection of crops, fertilizers, and cultural practices (Frissel, 1977). Most reports on the chemical quality of organic soils consider a three-compartment model as illustrated in Figure 6.1. Other studies may include water-soluble nutrients, and either crop uptake or nutrient leaching.

V. NITROGEN INDICATORS

A. Nitrogen Content and Fractions

Organic soils contain 5200 to 26,600 kg N ha^{-1}, averaging 14,400 kg N ha^{-1}, primarily as organic N, in the top 20 cm (Kaila, 1958a; Scheffer, 1976). As available

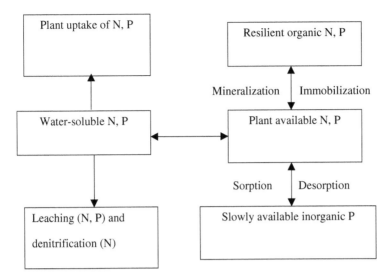

Figure 6.1   Model of nutrient transfer centered at the water-soluble form and involving processes assessed by biological and chemical indicators.

C pools are depleted in organic soils, the C/N ratio decreases, the ash content increases, and net N immobilization changes into net N mineralization (Tate, 1987). Part of the organic N reserve (3–20%) is tied up in the microbial biomass (Williams and Sparling, 1984). The enormous N supplying capacity of organic soils must be released at nonexcessive rates to minimize the risk of nutrient imbalance in crops, nitrate accumulation in the edible portion, and contamination of water and air by NO_x products.

The main N processes in soil–plant systems are mineralization of organic matter, ammonification, nitrification, denitrification, N immobilization by microbes, N uptake by plants, ammonium fixation or exchange, and ammonia volatilization. The rate of nitrate accumulation in peat materials is generally assessed as the rate of nitrification minus the rate of denitrification (Avnimelech, 1971), although significant N immobilization may reduce nitrate levels (Isirimah and Keeney, 1973). The simplest N turnover model assumes a single compartment for all organic N fractions, and first-order kinetics. More sophisticated models include two or more compartments (Jenkinson, 1990).

Nitrogenous carbon compounds in peat materials were characterized by Sowden et al. (1978) and related to N mineralization (Isirimah and Keeney, 1973). Incubation studies suggested that much of mineralizable N in priorly air-dried peat materials was derived from the acid-soluble organic N fraction and that considerable microbial turnover of soil N occurred (Isirimah and Keeney, 1973). Indicative products of N transformation were organic N, mineral N (NH_4-N and NO_3-N, sometimes NO_2-N), and N_2O.

Browder and Volk (1978) presented a model for organic soil subsidence linking the release of nitrate to CO_2 evolution. The multicompartmental biological submodel comprised a large compartment of nonliving carbon compounds decaying according to Michaelis–Menten kinetics, and a small compartment of active living carbon made of the microbial biomass and functioning extra-cellular enzymes. Nonliving carbon compounds were classified based on their degree of resistance to microbial break-down as difficultly hydrolyzable polysaccharides, insoluble aromatics, amino acids, soluble aromatics, and easily hydrolyzable aromatics. Nitrogen transformations involved available N, biomass N, and nonliving N in carbon compounds, as well as N fixation and denitrification.

B. Environmental Conditions for N Mineralization

Avnimelech et al. (1978) reported nitrate accumulation of 1000–2000 kg NO_3-N ha^{-1} yr^{-1} under field conditions in Israel. From bulk density, N content and subsidence rate, nitrification rate was estimated at 1400 kg NO_3-N ha^{-1} yr^{-1} in a Pahokee muck of the Florida Everglades (Tate, 1976), and 830 kg NO_3-N ha^{-1} yr^{-1} in New York organic soils (Duxbury and Peverly, 1978; Guthrie and Duxbury, 1978). Those figures were confirmed by laboratory experiments (Guthrie and Duxbury, 1978; Terry, 1980). The contribution of denitrification to nitrate removal is difficult to assess from N_2O determination alone, because the N_2 to N_2O ratio depends on the inhibitory effect of high nitrate concentrations on N_2O reduction to N_2, and to the adaptation capacity of the microbial community to high NO_3-N levels (Terry and Tate, 1980a). Fresh organic matter (e.g., root exudates and plant residues) stimulates denitrification in organic soils (Tate, 1976; Terry and Tate, 1980b). In addition to nitrification itself, the leaching of soluble organic N along with mineral N (Sahrawat, 1983), and N immobilization during incubation (Isirimah and Keeney, 1973) could affect the calculation of peat nitrification potential.

The main external factors influencing N transformation in organic soils are soil temperature and moisture (Browder and Volk, 1978). Nitrate peaked, and ammonium decreased, at 20°C in pristine peat materials (Kaila et al., 1953). Nitrification rate nearly doubled with a temperature increase from 24 to 36°C. As the temperature was raised between 30 and 60°C, nitrification rate decreased and denitrification rate increased (Avnimelech, 1971). Volk (1973) found that the rate of evolved C from Terra Ceia and Monteverde peats in Florida increased two- to threefold, depending on water table level, when temperature increased from 25 to 35°C. Microbiological processes have been shown to be most active at 35°C and 0.40 m^3 m^{-3} moisture content (Zimenko and Revinskaya, 1972). Waksman and Purvis (1932a) found the optimum range for fen peat decomposition to be between 50 and 80% on a fresh weight basis. Maximum carbon mineralization and nitrification in German organic soils occurred in the range of 60 to 80% of water retention capacity, while intensive denitrification occurred near or above water retention capacity (Scheffer, 1976). No significant effect of soil moisture tension on N mineralization was found within the range of 0.1 to 3.0 bar (10 to 300 kPa) in surface samples of a Pahokee muck in Florida (Terry, 1980). Optimum nitrate accumulation occurred near field capacity in a Hula peat in Israel (Avnimelech, 1971).

Lucas and Davis (1961) reported N content and pH as important soil factors that influenced N release and availability in organic soils. Acid peat materials (pH in water below 4.0) typically showed less than 1% N and C/N ratios around 60, while less acidic peat materials (pH in water above 5.0) had more than 2% N and C/N ratios near 30. The C/N ratio in Canadian organic soils was related to ash content and pH (MacLean et al., 1964): the C/N ratios varied between 15 and 20 in the 10–25% ash content range (i.e., above pH of 5.2 in water, or air-dried soil to a solution ratio of 1:4), and tended to stabilize at about 13–15 above 25% ash content. During the decomposition process, non-nitrogenous compounds decompose more rapidly than nitrogenous compounds, thus leading to an increase in the N content of residual peat (Waksman and Purvis, 1932a, 1932b; Murayama et al., 1990).

Carbon content is relatively constant in peat materials, therefore, the N content is a good indicator of mineral N mobilization (Puustjärvi, 1970). In Finland, hay yields without N fertilization were extremely low in organic soil materials containing less than 2% N (C/N ratio >30), and increased linearly up to 3% N. The zero-yield value corresponding to zero-N release was 1.7% N (Puustjärvi, 1970). The critical C/N ratio between N mobilization and immobilization, which depends primarily on available carbon, was 29 for pine grown in eutrophic *Sphagnum* woody peat materials (Puustjärvi, 1970). In forest peat materials (C/N ratio of 23.4) incubated at 30°C under aerobic conditions, rapid microbial activity was concomitant with a high degree of NH_4-N assimilation; under anaerobic conditions, NH_4-N accumulated due to limited microbial organic decomposition and low utilization of nitrogen (Williams, 1974).

As the pH in water falls below 5.0, the nitrification rate is markedly reduced. Nitrification in organic soils is performed by heterotrophic and autotrophic nitrifier populations (Tate, 1980a). The nitrifying bacteria were not found up to pH 5 in remoistened air-dry peat materials from the upper layer of a pristine peat material (Ivarson, 1977). Nitrification proceeded at a very slow rate in field-moist cultivated peat materials below pH 5.5 in water (Turk, 1943) and below 4.2–4.4 in $CaCl_2$ 0.1 *M* (Isirimah and Keeney, 1973; Scheffer, 1976); the pH in $CaCl_2$ 0.1 *M* is typically 0.5 to 1.1 pH units lower than pH in water (Isirimah and Keeney, 1973; Van Lierop and MacKenzie, 1977).

Incubation studies indicated that liming may increase ammonia volatilization and stimulate nitrification in pristine peat materials (Kaila et al., 1954; Kaila and Soini, 1957; Kivekäs and Kivinen, 1959). The peat samples were air-dried and ground before incubation, which increased by two- to fivefold the average amount of ammonium nitrogen, but affected nitrate to a small extent in peat materials (Kivekäs, 1958; Kaila, 1958a; Godefroy, 1977). A microbial flush can occur after remoistening air-dried soil (Ivarson, 1977). The Kivekäs and Kivinen (1959) results on 60 air-dry peat materials showed no clear relationship between pH and extracted ammonium and nitrate; however, compared with unlimed materials, nitrate content in limed peat materials increased by 32 ± 66 mg kg^{-1} after 1 month, and 149 ± 195 mg kg^{-1} after 3 months. In a long-term field experiment, Kaila and Ryti (1968) confirmed that liming an organic soil to reach pH 5.0 in $CaCl_2$ 0.2 N (1:2.5 soil to solution ratio) can double nitrate accumulation compared with pH 4.4 to 4.8, despite no effect on total mineral N accumulation.

C. Management Practices to Reduce Loads of Nitrate and Nitrous Oxide in Organic Soils

Due to intensive nitrification, nitrogenous fertilizers are rarely applied to organic soils in Florida (Terry, 1980). Sugarcane removes only 80–100 kg N ha^{-1} and the N loading rate to agricultural runoff water in the Everglades is 12 to 40 kg N ha^{-1} yr^{-1}, therefore, a large proportion of the 1000 to 1500 kg N ha^{-1} released annually must be denitrified (Terry, 1980). Thus, organic soils are environmentally at risk for nitrate and nitrous oxide pollution. Tate (1976) proposed decreasing the subsidence rate, thus decreasing the amount of inorganic nitrogen produced, and increasing denitrification by maintaining a high water table. The following management practices were implemented in Israel in order to prevent nitrate leakage into surface water (Avnimelech et al., 1978):

1. Improve the drainage system and reduce flooding
2. Maintain a relatively high water table (70 cm) during the summer to minimize subsidence and salt upward movement by capillarity
3. Lower water table before the rainy season to maximize storage and minimize seepage
4. Induce denitrification through the use of sprinkler irrigation
5. Select crops that reduce nitrate accumulation in the soil

Under cooler conditions, the N fertilization depends on soil moisture conditions and the crop being grown. In Indiana, N fertilization tended to minimize yield differences between the 40-cm water table and lower levels (60 to 100 cm) (Harris et al., 1962). In Ontario, the onion responded to N if irrigation was applied, because some N may be leached (Riekels, 1977). In New York, even if the amount of organic N mineralized annually is many times that required by the crops, vegetables grown in organic soils may respond to N fertilizers due to spring leaching events and insufficient rate of mineralization at low soil temperatures at the beginning of the growing season (Guthrie and Duxbury, 1978; Minotti and Stone, 1988). Nitrification inhibitors showed little mitigating effects in organic soils (Scheffer, 1976).

With provision for maintaining high crop quality during storage (Riekels, 1977), the N fertilization should be managed with parsimony in cool-region organic soils and according to crop requirements during the season, in order to reduce the accumulation of nitrate and nitrous oxide.

VI. PHOSPHORUS INDICATORS

A. Phosphorus Content and Fractions

Total P content averaged 1078 mg P kg^{-1} of air-dried soils (range: 375–1960) in pristine or cultivated Quebec organic soils (Parent et al., 1992), and 766 mg P kg^{-1} (range: 190–2350) in pristine Finnish organic soils (Kaila, 1956; Kaila, 1958b). Mean organic P proportions were 67% of total P (range: 41–86%) in Quebec organic soils and 78% (range: 55–95%) in Finnish organic soils. Total P accumulation in Finnish pristine peat materials was significantly correlated to the degree of humifi-cation, ash content, and N content (Kaila, 1956). Daughtrey et al. (1973a) found an average total P of 729 mg P kg^{-1} (range: 470–1185) in North Carolina organic soil materials containing more than 40% organic matter; organic P accounted for 75% of total P in the range of 72 to 80%.

Soil preparation may influence the organic P determination, but results are inconsistent. Soil drying may decrease available inorganic P content (Kivekäs, 1958), increase it (Daughtrey et al., 1973b; Godefroy, 1977), or show no definite trend (Anderson and Beverley, 1985). A closer examination of these results indicated a tendency for available inorganic P to decrease upon air-drying for soils below a water pH of 4.4 (air-dry materials), and for a P increase, or no change, for soils above pH 5.0. The P differences between field-moist and air-dry samples showed no trend in the 4.4–5.0 soil pH range. Compared with field-moist conditions, air-dried pristine peat retained more P, while air-dried drained peat (moorsh) released more P or showed no difference.

Fertilization at rates of 40 to 60 kg P ha^{-1} during more than 30 years increased markedly the amounts of total, organic, available (using a chemical extractant) and water-soluble P in a cultivated organic soil, compared with control or a low rate of 20 kg P ha^{-1} (Kaila and Missilä, 1956; Kaila, 1959a). Using the P-fractionation method of Wagar et al. (1986), Sasseville (1991) found that more scarcely available resin-P accounted for 13% of total P, easily available NaHCO$_3$-P for 18%, more scarcely available NaOH-P for 30%, and residual P in humus for 39%, in Quebec moorsh soils. The pH of air-dry soil was determined in 0.01 M CaCl$_2$ using a 1:2 soil to solution ratio. In a newly reclaimed Sphagno–Fibrisols, total P increased linearly from 604 to 978 and 1170 mg P kg^{-1} with lime additions of 0 (pH = 4.2), 6 (pH = 5.1), and 12 (pH = 6.2) Mg ha^{-1}, respectively (Parent et al., 1992). A considerable increase in total P occurred above pH 5.8, in residual P above pH 5, and in the inorganic NaHCO$_3$-P and NaOH-P fractions above pH 5.5. The proportions of total P decreased markedly at pH values exceeding 4.5 for both organic and inorganic NaHCO$_3$-P, and 6.2 for resin P.

B. Environmental Conditions for Release of Organic P

The dynamics of organic P in organic soil materials have been studied, assuming a single compartment for soil organic P. Apparently, more than half of the P added to an acid (pH 4.6) cultivated organic soil over 30 years has been converted to organic P; newly reclaimed and fertilized organic soils accumulated 74 mg organic P kg^{-1} and 136 mg inorganic P kg^{-1} after 4 years under cultivation (Kaila and Missilä, 1956). Kaila (1958b) found that organic P decreased by 5 to 15%, and inorganic P increased correspondingly, in acid pristine peat materials incubated for 4 months at 27°C. The pH in water using a soil to solution ratio of 1:4 varied between 4.0 and 5.6. In incubation experiments, organic P mineralization was found to be in the range of 2.2 to 20% of total soil P (38 to 185 kg P ha^{-1} yr^{-1}) for central Florida organic soils and of 0.8 to 1.1% of total P (16 to 23 kg P ha^{-1} yr^{-1}) for southern Florida

(Reddy, 1983). In southern Florida organic soils, P mineralization rate ranged from 6 to 72 kg P ha^{-1} yr^{-1} for drained soils and from 36 to 88 kg P ha^{-1} yr^{-1} for flooded soils (Diaz et al., 1993). All extractable forms of inorganic P in peat samples incubated anaerobically were greater than or about equal to P concentrations in samples incubated aerobically, and were thought to derive mainly from organic P (Racz, 1979).

Daughtrey et al. (1973a) found that organic soil materials containing more than 380 mg P kg^{-1} (organic C/P ratio < 560) had mineralization rates 4 times greater than soils containing less than 230 kg P kg^{-1} (organic C/P ratio > 1000). Those results differ from the critical organic C/P ratio of 300 proposed by Stevenson (1986) for balance between organic P mineralization and immobilization in mineral soils. The rapid rate of organic P mineralization in organic soils indicated that mineralization of organic P was independent of inorganic P content, and that immobilization of organic P was not necessarily rapid when inorganic P level was quite high (Daughtrey et al., 1973a). In acid organic soil materials, biomass P can account for 7 to 22% of total P, thus contributing substantially to the P turnover (Williams and Sparling, 1984); however, microbial P mobilization in organic soils has been neglected compared with the numerous experiments relating sorption of inorganic P to Ca, Al, and Fe compounds.

C. Environmental Conditions for Release of Inorganic P

Direct evidence of P sorption by mineral matter was obtained by adding sorptive materials to the soil. Droughty (1930) saturated a peat material with calcium, ferric and aluminium chlorides. Calcium had little effect on P fixation due to the peat interfering with the precipitation of calcium phosphates up to pH 6.7; Fe and Al fixed P effectively in the pH range recommended for agricultural crops. These results were confirmed later by Larsen et al. (1959), Fox and Kamprath (1971), and Bloom (1981), who added Al to organic soils or to low P-fixing, high organic matter, mineral soils. It has been suggested that inorganic P is loosely held by cations on organic colloids (Daughtrey et al., 1973a), due in part to the formation of insoluble Al-organic matter complexes (Clark and Nichol, 1966). Salonen et al. (1973) found that oats grown in *Sphagnum* peat that was poor in phosphate-fixing substances required substantially less fertilizer P (1/8) to attain maximum yield compared to a gyttja clay.

Indirect evidence of P sorption by mineral matter in organic soils was obtained from correlation studies. The P sorption in newly reclaimed organic soils was closely related to Fe (Kaila and Missilä, 1956), Al (Kaila, 1959b), total sesquioxide content (Larsen et al., 1958, 1959; Miller, 1979), as well as pH, Ca content, and ash content (Porter and Sanchez, 1992). Native calcium carbonate likely regulated P sorption in alkaline organic soils (Richardson and Vaithiyanathan, 1995), while recently limed organic soils showed reduced P availability (Lawton and Davis, 1956; Okruszko et al., 1962; Hamilton and Bernier, 1973).

In mineral soils, P saturation indices were developed from routine soil analyses using extractable P as the numerator and reactive Al and Fe as the denominator (Van der Zee et al., 1987; Khiari et al., 2000). A rapid method should also be developed for determining the P fixation capacity of organic soils. Porter and Sanchez (1992) found attractive the use of ash content as indicator of the P sorption capacity in Florida organic soils due to its ease of determination. Miller (1979) suggested using total Fe and Al for Ontario organic soils; however, these analyses are not conducted routinely in soil testing laboratories.

Harris and Warren (1962) found that Fe and Al alone were not reliable P fixation indicators when previous fertilization satisfied their fixation capacities. Thus, an Al–Fe index of P fixation must include a measure of soil reactive P. The P sorption by soils relates solution P concentration to solid-phase P at equilibrium. The Freundlich equation is often used to describe that relationship as $x/m = kC^{1/n}$, where x/m is solid-phase P (mg P kg^{-1} of dry soil), C is solution P concentration (mg P L^{-1}), k is a constant related to P fixation capacity (L kg^{-1}), and $1/n$ is a unitless constant related to P fixation intensity.

Using the ratio of exchangeable P to Freundlich k, Kaila (1959a) found that high rates of P fertilization during more than 30 years increased k and relative P content of an organic soil. Kaila (1959b) determined the Freundlich k constant, an indicator of P sorption, and extracted Fe and Al with 0.1 M HCl, and exchangeable P with 0.1 M KOH-K$_2$CO$_3$, in 134 organic soil materials. The Al and Fe can be added up on a molar basis after dividing Fe and Al by their respective molecular weights (i.e., 27 for Al and 56 for Fe). The 100P/(Al + Fe) molar ratio can then be assessed as an indicator of the inorganic P saturation in a given organic soil, as was done for mineral soils using other extraction methods (Van der Zee et al., 1987; Khiari et al., 2000). After removing two outsiders (soils no. 26 and no. 91), one containing almost no Fe and Al, the results are presented graphically in Figure 6.2. Despite the limitations of using Freundlich k as a P sorption index, such as high correlation (r

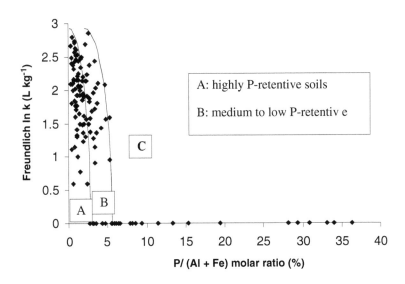

Figure 6.2   Relationship between P sorption and the P/(Al + Fe) ratio in the 0.1 M HCl extract. (Adapted from Kaila, A. 1959b. *J. Sci. Agric. Soc. Finland*, 31:215–225. With permission.)

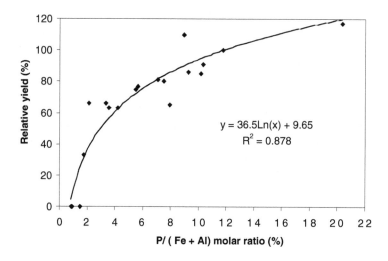

Figure 6.3 Relationship between percentage yield of Ladino clover and the P/(Al + Fe) ratio in the 0.5 *M* HCl extract. (Adapted from Okruszko, H., Wilcox, G.E., and Warren, G.F. 1962. *Soil Sci. Soc. Am. Proc.,* 26:71–74. With permission.)

>0.9) between parameters in nonlinear models (Robinson, 1985) as shown by the Freundlich data of Porter and Sanchez (1992), the k–P saturation index relationship was informative. Obviously, the P retention capacity (ln (k)) decreased rapidly as P saturation increased (Figure 6.2). Three soil groups can be recognized with approximative ranges of P saturation index of 0 to 1, 1 to 2, and more than 2. Over a P saturation index of 2, chemical P retention capacity was completely lost.

The agronomic value of the 100P/(Al + Fe) molar ratio was examined using the Ladino clover data published by Okruszko et al. (1962). The coefficient of determination was surprisingly high for the first cut (Figure 6.3), and 0.85 for the three cuts altogether. Thus, analytical methods for P, Al, and Fe in organic soils can be routinely conducted in soil testing laboratories, and a P saturation index can be computed using extractable P as numerator and extractable Al and Fe as denominator. Soil testing procedures can thus be implemented not only to make P fertilizer recommendations, but also to assess the environmental risk of the fertilized soil–plant systems.

D. Good Management Practices to Reduce P Load

In the Florida Everglades, potential areas to reduce P in drainage water include the following best management practices (BMPs) (Izuno et al., 1991):

1. **Improved management of drainage rate, volume, and timing** — Reduced-P loading must be obtained from reduction in P concentrations during large volume discharges instead of from farm drainage retention only. Fallow flooding to control pests increases P load in surface waters; however, the risk of crop flooding by reduced drainage during the rainy season, and the timing of fallow flooding periods, limit the application of those BMPs.

2. **Fertilizer reduction and enhanced crop rotation strategies, considering locations of higher fertilizer rates and the period of time that the land had been intensively farmed** — To reduce P concentrations while obtaining efficient use of plant available P, BMPs may include fertilizer form and placement, high-yielding and P-mining crops grown in rotation, variable-rate P fertilization depending on spatial variability of soil properties and seasonal crop requirements, and improved soil testing accounting for organic P mineralization.

Mineral materials rich in Fe and Al may also be mixed with organic soils to reduce P leaching (Scheffer and Kuntze, 1989). Cogger and Duxbury (1984) suggested to use Fe and Al levels to predict P leaching losses in New York organic soils, to consider mineralized P, and to reduce fertilizer use without sacrificing yield and quality where crop response to P is small. Indeed, no loss occurred in the yield or quality of onions grown continuously for 3–4 years or more in New York P-rich organic soils without P fertilization (Minotti and Stone, 1988). Similarly, Tremblay and Parent (1989) found that the carrot responded more to the past than present fertilization regime. Bélair and Parent (1996) found that crop rotation involving a nonfertilized cereal fallow, preceding or following onion crops, could maintain a highly productive carrot crop during two consecutive years without nematicides in an organic soil already infested by the root-knot nematode. High yielding crops are effective sinks for soil P.

Soil testing, a prerequisite for implementing BMPs, is complicated by the larger range of bulk densities and water contents in samples of organic soils compared with mineral soils, and with a greater influence of organic soil drying on soil chemical attributes. In soil testing circles, it has been agreed that volumetric expressions for nutrients (w/v) should be used. The significance of soil drying, which can be tracked back to Kaila's work in Finland, is still debated. Van Lierop (1981) proposed a reconstitution of the bulk density (BD) of field-moist samples in 17-mL plastic cups before proceeding to extraction, because drying increased BD about twofold in Quebec organic soils. Anderson and Beverley (1985) concluded that Florida organic soils should undergo a standardized drying procedure in order to bring soil moisture into equilibrium with preparation procedures. Daughtrey et al. (1973a, b) were strong proponents of volumetric determinations of air-dried soils, because the drying process led to some mineralization of organic P in North Carolina organic soils. This mineralization was also reflected by increases in Mehlich-1-extractable P. Parent et al. (1997) followed the seasonal fluctuations of nutrient status in a Quebec organic soil, using field-moist samples and the saturated paste method. They found that using nutrient multiratios instead of nutrient concentrations reduced the influence of operators on soil volume, which caused variations in nutrient concentrations.

VII. RELATED C, N, AND P CYCLES

The drainage and agricultural use of organic soils transforms the upper peat material into terrestrial humus forms (Pons, 1960; Van Heuveln et al., 1960) or moorsh (Okruszko and Ilnicki, Chap. 1, this volume) across a plethora of physical, chemical, and biological processes. This results in soil subsidence that is related to the water table level (Neller, 1944). *Nitrosomonas* and *Nitrobacter* as well as catabolic activities increase sharply in peat materials (Herlihy, 1972) down to the water table (Tate, 1979). Microbial activity is stimulated by O_2, deeper soil and increased pH upon liming (Tate, 1979, 1980b, 1980c). Greater enzyme activities contribute to organic matter oxidation (Tate, 1984; Duxbury and Tate, 1981; Mangler and Tate, 1982; Tate, 1984). Thus, enzyme inactivation (Mathur and Sanderson, 1980) or mitigation measures that limit microbial metabolism of a wide variety of carbon sources in soils (Tate, 1984) have been suggested to reduce organic soil subsidence rate.

Computations based on changes in bulk density and ash content, as well as CO_2 evolution with soil depth, supported the concept that biological oxidation is the single most important factor in long-term organic soil subsidence. From long-time subsidence records as well as soil bulk density, ash content, and N supply to crops, Schothorst (1977) attributed 85% of organic soil subsidence in polders of western Netherlands to biological oxidation. Combining CO_2 evolution data with the USDA subsidence line elevation loss measurements in Florida, Volk (1973) calculated that biological oxidation accounted for 73% of total subsidence in a Monteverde muck, and for 58% of total subsidence of a Terra Ceia muck.

Scaling subsidence survey and research data using a limnic borohemist of the Sainte-Clotilde-de-Châteauguay federal experimental farm as reference soil, Parent et al. (1991) estimated long-term subsidence rate in southwestern Quebec. The subsidence equation was in the form of a second-order kinetics as follows:

$$d = \frac{1}{(0.5936 + 0.0112t)} \tag{6.1}$$

where d is residual peat thickness in cm and t is time in years. Equation 6.1 is valid for up to 50 years of vegetable production, which is the computed half-live of the organic soil, and an initial peat thickness of 1.7 m (1/0.5936). From Equation 6.1, the subsidence rate reaches 1.6 cm yr^{-1} in the long run. In the same soil (C/N ratio of 15–16), Campbell and Frascarelli (1981) calculated from CO_2 evolution monitoring that biological oxidation was 0.8 cm yr^{-1}. In the same area, but with peat materials showing a C/N ratio of 33, Macmillan (1976) obtained a net loss of 17,160 mg CO_2-C kg^{-1} (i.e., approximately 17,000 kg ha^{-1}) or 0.6 cm for an arable layer (0–20 cm) with a bulk density (BD) of 0.28 g cm^{-3} [the average BD value for Quebec moorsh soils is 0.28 ± 0.08 g cm^{-3} for soils containing 0.20 ± 0.10 kg ash kg^{-1} (Parent et al., 1991)]. Thus, biological oxidation accounted for approximately 40–50% of the long-term subsidence in that organic soil of southwestern Quebec. Local soil deflation due to water erosion followed by land leveling (Parent et al., 1982), as well as lateral oozing of the calcareous limnic materials underneath, probably contributed to soil subsidence to a greater extent in this soil than in the Dutch or Florida soils.

According to McGill and Cole (1981), reduced C is the main energy supply for internal soil nutrient cycling, primarily N, S, and organic P, thus providing a framework for interactive nutrient turnovers. For mineral soils, Stevenson (1986) proposed that, above a C/P ratio of 300, the soil would give up P to the microbes; below 200, a net mineralization of organic P would occur. Assuming stoichiometric relationships among nutrients in a stabilized material, N, P, and S must be released concomitantly as waste products during microbial oxidation of C to CO_2. Examining literature data on soil nutrient ratios, however, McGill and Cole (1981) proposed that C, N, and C-bonded S are mineralized together as driven by the biological search for energy, while organic P and sulfate esters are mineralized independently through enzymatic catalysis (phosphatase and sulfatase activities, respectively) external to the cell membrane, as controlled by nutrient requirements. As a result, the mobilization processes in soils would be nutrient-specific, and the P and S supply to plants from soil organic matter (SOM) would not simply follow proportionality laws. In that model, C was the driving factor, but P had ultimate control on organic matter cycling and accumulation in soils.

In another approach, Damman (1988) proposed that the critical C/N ratio for N mineralization must be an exponential function of the microbial N requirement. The critical C/N ratio of 14–15 must be reached for meeting 100% of the microbial N requirements in a given mire ecosystem. The critical C/N ratio must increase at the suboptimal microbial N requirement, as controlled presumably by P shortage in mire ecosystems. The P shortage promotes phosphatase activities. As shown by Parent and MacKenzie (1985), phosphatase activity is higher in pristine compared with cultivated and fertilized moorsh materials, indicating biological response to P-deficiency. Therefore, stoichiometric relationships can be obtained only if they are close to optimum conditions for meeting 100% of the microbial N and P requirements. Under suboptimum conditions, N and P mineralization proceeds following an exponential rather than a stoichiometric relationship with microbial requirements.

Assuming stoichiometric C/N relationships and an optimum microbial N requirement, the C and N mineralization rates can be associated with subsidence rates of organic soils. Nesterenko (1976) reported the release of 450 kg N ha^{-1} and net loss of 100 kg N ha^{-1} from 17,000 kg ha^{-1} of consumed organic matter per year in a Belarus organic soil near Minsk (assuming a C/N ratio of 22 and 58% C in SOM). Schothorst (1977) estimated that 12,000 kg ha^{-1} of SOM or 0.6 cm yr^{-1} of biological oxidation would release approximately 480 kg N ha^{-1} in a Dutch organic soil (assuming 4% N in SOM, thus a C/N ratio near 15, and 50% N-use efficiency by harvested crops). In a Pahokee muck showing 45% C, a C/N ratio of 23.6, and a bulk density of 0.25 g cm^{-3}, Tate (1980b) calculated that for each cm yr^{-1} of organic soil subsidence, 11,000 kg C ha^{-1} and 11,000/23.6 = 466 kg N ha^{-1} must be released. Those computations agree with N uptake by agronomic crops grown in organic soils in The Netherlands (Schothorst, 1977) and Poland (Gotkiewicz, 1977). Because no crop grown in the Everglades could take up 1400 kg N ha^{-1} for a subsidence rate of 3 cm yr^{-1}, Tate (1980b) concluded that most of the mineralized N must be denitrified (Terry and Tate, 1980b) or leached into waterways (Hortenstine and Forbes, 1972, Volk and Sartain, 1976). Denitrification is a major sink for nitrate in drained organic soils (Terry and Tate, 1980b; Martikainen et al., 1995), thus indicating the presence of large amounts of easily available C in the soils.

Mire classification is based primarily on water chemistry (acidity as well as cation, bicarbonate and sulfate concentrations), therefore, little documentation exists about the role of N and P in mire building. Koch-Rose et al. (1994) related the trophic level of Everglades organic soils to the N/P ratio in pore water: eutrophic (fen) peats showed N/P ratios of about 3, compared to 39–551 for oligotrophic (bog) peats. The N/P ratio is often not balanced in peat materials (100 to 2–4) compared with the tree biomass grown on cutover peatlands (100 to 10–13) (Kaunisto and Aro, 1999). Verhoeven et al. (1988) found that it is direct N and P supply from a ditch instead of mineralization of organic N and P from peat that explained the composition and productivity of mire species in a Dutch freshwater mire. However, nutrient uptake problems for plants may be encountered in extremely acidic mire ecosystems (Waughman, 1980). Also, greater immobilization of inorganic P into the microbial biomass can occur in organic soils of low P availability, and foliar nutrient reabsorption from senescing to fresh foliage can be a predominant pathway for P recycling independent of soil P availability (Walbridge, 1991). DeBusk et al. (1994) established a spatial link between agricultural and urban uses of organic soils, increased P load and storage, eutrophication, and the occurrence of cattails (*Typha domingensis* Pers. and *T. latifolia* L.) in the Everglades.

Waksman and Purvis (1932b) concluded that microbial growth in previously dried peat followed by incubation was limited by oxidizable carbon rather than NO_3-N or P shortage. Bridgman and Richardson (1992) confirmed those views for incubated field-moist peats. Because nitrogenous compounds decompose rapidly in *Sphagnum* plants, while cellulose and hemicellulose remain recalcitrant, there is little microbial synthesis and, therefore, net N mineralization as NH_4-N occurs (Waksman and Purvis, 1932b). Conversely, under suboptimal microbial activity, net N mineralization can occur at high C/N ratios if severe constraints on the size and activity of the microbial pool are imposed, as may be imposed by P shortage instead of a lack of easily available C in *Sphagnum* peats (Damman, 1988; Williams and Silcock, 1997, 2001). Kong and Dommergues (1970, 1972, 1973) showed that cellulolytic activity was limited by P in calcitic peats and that carboxymethylcellulase activity was restricted by both N and P in acid peats.

The contradictory results on rate-limiting nutrients presented previously is attributable to the fact that, in short-term laboratory incubation experiments, any change in the enzymatic or microbial make-up driven by N or P treatments had little observable effect on CO_2 production. This would also be the case for longer-term incubations, where the enzymatic make-up and microbial populations must change with time and affect CO_2 production, or under field conditions where soil genesis builds up over the years in drained peat materials (Van Heuveln et al., 1960).

Superphosphate additions increased mineral N content and stimulated nitrification in the upper 50 cm layer of an organic soil cultivated for 35 years (Kaila, 1958a). The N to organic P ratio varied between 12 and 133 in pristine peat materials, and between 8 and 74 in moorsh soils, which was much higher than the ratios of 5 to 16 found in mineral soils (Kaila, 1956). Brake et al. (1999) observed a decrease in the C/N ratio from 39 to 22 and in the C/P ratio from 1928 to 457 in the oxic horizon of cultivated and fertilized organic soils compared with the anoxic layers. Because biomass P made up 32% of total P, the soil microbes dominated the turnover of P in those organic soils. Thus, one should establish environmentally acceptable amounts of total and biomass P, as they relate to cultural practices and soil properties, in organic soils for assessing the risk of P pollution.

VIII. NEED FOR N AND P INDICATORS

Soil quality indicators that can be acquired from routine analyses are necessary to improve soil management. First, we need a P saturation index that is highly correlated to P leaching potential and yield of agronomic crops. This is possible using an extraction procedure that will provide P, Al, and Fe concentrations of diagnostic value. Second, we need an indication of the potential N and P mineralization in specific fractions of soil organic matter. These fractions should vary with intensity of the moorshing process and crop rotation. In order to control soil quality, we could measure biomass C, N, and P supporting nutrient cycling and biodegradation functions, denitrification enzyme activity as an index of denitrification capacity, organic N and P mineralization as important contributors to soil fertility, and soil respiration as overall biological activity (Groffman et al., 1996).

Ratios between C, N, P, and S have also been proposed to assess the mineralization-immobilization processes in soils (Walker and Adams, 1958; Parton et al., 1988). Dual ratios can be easily computed. Higher-level ratios may cause greater operational problems, such as using the (C/N)/P or C/(N/P) ratio computation procedures, and so on. The use of multiratios, derived from the compositional data analysis (Aitchison, 1986), should assist in solving those problems. The computations require a filling value, R, between 100% and analytical results, because all components must add up to 100% in a closed system. Knowing C, N, and P expressed as percentages, and assuming one compartment each for C, N, and P, we compute R as follows:

$$R = 100 - C - N - P \qquad (6.2)$$

The geometric mean of the simplex is computed as follows:

$$G = (C \times N \times P \times R)^{0.25} \qquad (6.3)$$

Each component is transformed into a multiratio as follows for N:

$$V_N = \ln(N / G) \qquad (6.4)$$

and similarly for C, P, and the filling value if required. The multiratio is scale-invariant and can provide useful indicators of potential C, N, and P mineralization or immobilization in a specific compartment of soil organic matter (Parent et al., 2000). The C/N, C/P, and N/P dual ratios are included in multiratios. The multiratios may be expanded to other elements like S (Parent et al., 2000). In addition, single-element compartments can be split into subcompartments or subsimplexes, and

Table 6.6   Properties of the Arable Layer of Acid Organic Soils and Organic Epipedons from North Carolina

No.	C/N (Unitless)	C/P (Unitless)	N/P (Unitless)	V_N (Unitless)	V_P (Unitless)	Mineralized P (mg P kg^{-1})	Dilute Acid Extractable P[a] (mg P kg^{-1})
4	33	744	23	−1.43	−4.55	33	18
5	33	392	12	−1.30	−3.77	103	71
6	48	1394	29	−1.09	−4.46	27	23
7	32	529	16	−1.14	−3.93	55	85
8	28	263	9	−1.58	−3.81	20	182
9	41	1528	37	−0.97	−4.58	31	22
10	39	1500	38	−0.89	−4.54	11	48
11	36	467	13	−1.43	−4.00	45	19
12	29	430	15	−1.25	−3.97	82	37
13	33	1040	32	−1.05	−4.51	34	10
14	34	576	17	−1.32	−4.17	112	50
15	40	640	16	−1.28	−4.06	72	33
16	29	465	16	−1.41	−4.18	74	55

[a]Double acid method (Daughtrey et al., 1973a).

Source: Applied to data from Daughtrey, Z.W., Gilliam, J.W., and Kamprath, E.J. 1973a. *Soil Sci.*, 115:18–24.

analyzed as subcompositional datasets as described in detail by Aitchison (1986). Examples of subsimplexes are C fractions (heavy, light, and soluble organic matter), N fractions (insoluble or soluble organic N, mineral N as NH_4-N, NO_3-N, and NO_2-N) and P fractions using P fractionation procedures.

A V_P critical value can be derived from the Daughtrey et al. (1973a) dataset on organic P mineralization. The authors converted results from lb/acre to mg P kg^{-1} using bulk density values for an arable layer of 900 cubic yards (0–17 cm), then to percentages relative to maximum P mineralization. Transformed data are presented in Table 6.6.

A close examination of the data indicated abrupt gradients between three classes of P mineralization potential (Figure 6.4): from left to right, low values are shown, then a cluster of high values, followed by a tail of low values. The critical values computed as median values between adjacent P-mineralization classes on the right-hand side were 672 for the C/P ratio (Figure 6.4) and 20 for the N/P ratio (Figure 6.5). Interestingly, the single point on the left side of the diagram (Figures 6.4 and 6.5), having highest amount of double-acid available P (soil no. 8 in Table 6.6), showed low P mineralization potential despite a C/P ratio of 263 and a N/P ratio of 9; presumably, the biological demand, thus phosphatase production, was low due to very high availability of inorganic P in the soil.

Putting C, N, and P together as compositional data, the P mineralization first increased linearly with V_P, then showed no significant trend (Figure 6.6). The P mineralization potential remained low at V_P values under −4.4, then reached a maximum at V_P = −4.2. For V_P values exceeding −4.1, P mineralization either decreased, or was maintained, as probably related to the intensity of biological demand and phosphatase activities. This phenomenon can be explained by the McGill and Cole (1981) dichotomous concept for controlling P mineralization:

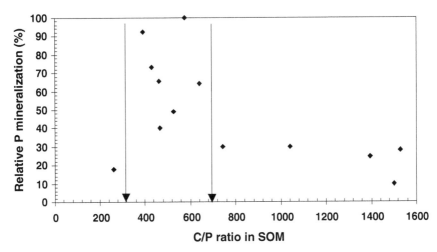

Figure 6.4   Relationship between organic P mineralization and the C/P ratio in soil organic matter (SOM) of acid peat materials. (Original data from Daughtrey, Z.W., Gilliam, J.W., and Kamprath, E.J. 1973a. *Soil Sci.*, 115:18–24. With permission.)

microbial requirements dominate for V_P values below –4.3, and biochemical or phosphatase-driven hydrolysis of organic P predominates for V_P values above –4.1. The Damman (1988) theory and the Kong and Dommergues (1970, 1972, 1973) findings on P shortage controlling microbial requirements in mires also holds for V_P values below –4.3. The compositional data analysis as conducted using Eqs. 6.1 – 6.3 was in line with present concepts on organic P transformation pathways in soils.

In typical moorsh soils of southwestern Quebec, the C/N ratio tends to stabilize near 15, with a typical C/P ratio of 340 in SOM (Parent et al., 2000). Because 340 < 672, the P should be mineralized at a high rate. Assuming a single pool of humic organic matter in the moorsh horizon and 580 g C kg^{-1} in SOM, SOM must contain 38.7 g N kg^{-1} and 1.7 g P kg^{-1}. Because $R = 379.6$ g kg^{-1} and $G = 61.7$, then $V_P = -3.6$. The computed V_P exceeded the critical V_P of –4.3 shown in Figure 6.6. Thus, high or low P mineralization is expected to occur in a typical Quebec moorsh soil, depending on biological demand controlling phosphatase activity. Presumably, such process contributed to the nonresponse of vegetable crops to the P fertilization observed in the Quebec fertilizer trial described earlier, and in the New York fertilizer trial of Minotti and Stone (1988).

On the other hand, little information is available on the relationship between N mineralization potential with both soil N and organic P in organic soils. Because the typical C/N ratio of 15 is less than the critical ratio of 29 (Puustjärvi, 1970) for organic soils, N should mineralize at high rate. Assuming a single SOM pool in the moorsh horizon and 580 g C kg^{-1} in SOM, with a critical C/N ratio of 29 and a critical C/P ratio of 672, SOM contains 20 g N kg^{-1} and 0.86 g P kg^{-1}. Because $R = 399.14$ g kg^{-1} and $G = 44.7$, critical $V_N = -0.8$. For a C/N ratio of 15 and a C/P ratio of 340 for a typical moorsh soil, $V_N = -0.5$, which is higher than the assumed critical V_N of –0.8. Obviously, net N mineralization must occur in those soils, as indicated by the nonresponse of vegetable crops to the N fertilization observed in

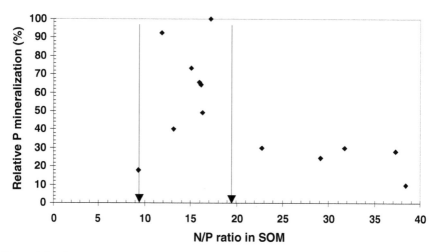

Figure 6.5 Relationship between organic P mineralization and the N/P ratio in soil organic matter (SOM) of acid peat materials. (Original data from Daughtrey, Z.W., Gilliam, J.W., and Kamprath, E.J. 1973a. *Soil Sci.*, 115:18–24. With permission.)

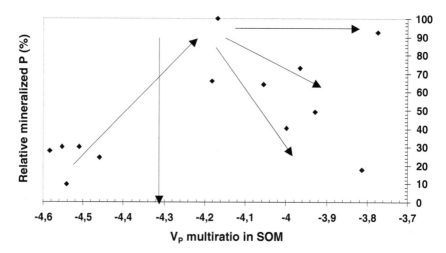

Figure 6.6 Relationship between organic P mineralization and V_P ratio in organic matter of acid peat materials. (Original data from Daughtrey, Z.W., Gilliam, J.W., and Kamprath, E.J. 1973a. *Soil Sci.*, 115:18–24. With permission.)

the Quebec fertilizer trial. For N-mineralizing moorsh soils, a constant C/N ratio value and organic soil subsidence rate can be used to assess yearly N mineralization (Tate, 1987). For a Quebec soil with a biological contribution to subsidence of 0.8 cm yr^{-1}, 20% ash, bulk density of 0.25 g cm^{-3}, and a C/N ratio of 15, the soil must release up to 620 kg N ha^{-1} yr^{-1} across a 0–30 cm moorsh layer. Such amounts largely exceed crop requirements. Thus, C, N, and P multiratios may be instrumental in predicting the C, N, and P mineralization rates as related to subsidence and the moorshing process.

The V_C, V_N and V_P (as well as V_R) indicators can be diagnosed either using critical ranges, or as standardized compositional variables using a chi-squared (χ^2) holistic approach (Khiari et al., 2001a, 2001b, 2001c). The sum of n squared standardized independent variables is distributed like a χ^2 variable with n degrees of freedom (Ross, 1987). Therefore, we need the respective mean and standard deviations from a target population to standardize the V_C, V_N, V_P, and V_R indicators as follows for V_P:

$$I_P = (V_P - V_P^*) / s_P^* \tag{6.5}$$

where I_P is the standardized phosphorus index, V_P is the phosphorus multiratio for the specimen under diagnosis, and V_P^* and s_P^* are the mean and standard deviation of V_P values from a target population. For example, for the population of data presented in Table 6.6, the V_P^* value is –4.052 and s_P^* is 0.105. For a V_P value of –3.6, the P index is computed as follows:

$$I_P = (-3.6 + 4.052) / 0.105 = 4.3 \tag{6.6}$$

A positive I_P value means relative P excess.

The global imbalance index could be computed as a χ^2 value as follows:

$$\chi^2 = I_C^2 + I_N^2 + I_P^2 + I_R^2 \tag{6.7}$$

Defining a critical χ^2 value to diagnose the result of Equation 6.7, a global picture is provided for a complex system involving multiple interactions between C, N, P, and the undetermined components R (e.g., S, H, O, ash). The critical χ^2 value, as well as V_P^* and s_P^* values and other values for standardization, can be defined according to management objectives (retarding subsidence or the moorshing process, increasing SOM mineralization, reducing fertilizer applications, altering nutrient cycling in the ecosystem, ...). As the critical χ^2 value is reduced, ranges of I_C, I_N, I_P and I_R, as computed from the square root of I_C^2, I_N^2, I_P^2, and I_R^2, respectively, must be further squeezed to reach the new target of χ^2 value.

If, for example, a critical χ^2 value of 4 is selected for the previous population as a target to improve the control of the P pollution, subsidence rate, and nitrification, and we obtain an I_P value of 4.3 as in Equation 6.5, I_P^2 is therefore 18.7 and the balance objective (the χ^2 value must not exceed 4) that was fixed for the system cannot be met; this is due to the fact that too much P would be mineralized from the soil. As a result, a decision to reduce sharply the P fertilization must be made. The same holds for N. Results of N and P fertilizer trials in Quebec and New York indicated that the P fertilization of moorsh soils can be avoided for many years without any agronomic loss of high-value crops; however, the C index could not be adjusted as rapidly or economically compared with N or P.

It is suggested that both V_N and V_P indicators be diagnosed in combination with routine soil tests on inorganic P or soil P saturation as well as soil acidity to improve the management of cultivated organic soils. Nutrient ratios in living plants, especially N/P, are often used to diagnose the nutrient status of mire plants (e.g., Walbridge, 1991). Foliage analysis and Compositional Nutrient Diagnosis of the P or N status of the agroecosystem computed similarly to Eqs. 6.1–6.3 (Khiari et al., 2001a) can also be useful as complementary tools for managing N and P in agronomic crops (Khiari et al., 2001b, 2001c) grown in organic soils.

IX. CONCLUSION

Sustainable moorsh soil management should move from input-based to knowledge-based systems using BMPs. In order to implement BMPs and convince the users with scientific information showing that such BMPs are profitable or at least safely applied with no financial loss, we need a reliable decision-support procedure. To achieve that goal, new indicators are necessary for N and P in organic soils. We proposed that P saturation index based on routine soil analysis for P, Al, and Fe in acid peats (e.g., the Mehlich III method currently used in Quebec) or for P, Ca, and carbonates in calcareous peats, as well as N and P turnover attributes of SOM (e.g., V_N, V_P) based on agroecosystem characteristics (i.e., relative to specific crop–soil–water features), should be developed and tested.

ACKNOWLEDGMENTS

This project was supported by the Natural Sciences and Engineering Research Council of Canada (OG no. 2254).

REFERENCES

Agriculture Canada. 1992. *The Canadian System of Soil Classification* (in French). Publ. 1646, 2nd ed., Research Branch, Agriculture and Agri-Food Canada, Ottawa.

Aitchison, J. 1986. *Statistical Analysis of Compositional Data.* Chapman & Hall, New York.

Anderson, D.L. and Beverley, R.B. 1985. The effects of drying upon extractable phosphorus, potassium and bulk density of organic and mineral soils of the Everglades. *Soil Sci. Soc. Am. J.,* 49:362–366.

Asselin, M. 1997. Computer model for rational use of fertilizers and the correction of nutrient imbalances in vegetable crops on organic soils (in French). Canada-Quebec Agreement Sustain. Environ. Agric. Rep. 13–67130811–046, Quebec.

Avnimelech, Y. 1971. Nitrate transformation in peat. *Soil Sci.,* 111:113–118.

Avnimelech, Y. et al. 1978. Prevention of nitrate leakage from the Hula basin, Israel: a case study in watershed management. *Soil Sci.,* 125:233–239.

Bélair, G. and Parent, L.E. 1996. Using crop rotation to control *Meloidogyne hapla* Chitwood and improve marketable carrot yield. *Hortscience,* 31:106–108.

Bloom, P.R. 1981. Phosphorus adsorption by an aluminium-peat complex. *Soil Sci. Soc. Am. J.,* 45:267–272.

Brake, M., Höper, H., and Joergensen, R.G. 1999. Land use-induced changes in activity and biomass of microorganisms in raised bog peats at different depths. *Soil Biol. Biochem.*, 31:1489–1497.

Bridgman, S.D. and Richardson, C.J. 1992. Mechanisms controlling soil respiration (CO_2 and CH_4) in southern peatlands. *Soil Biol. Biochem.*, 24:1089–1099.

Browder, J.A. and Volk, B.G. 1978. Systems model of carbon transformations in soil subsidence. *Ecol. Model.*, 5:269–292.

Campbell, J.A. and Frascarelli, L. 1981. Measurement of CO_2 evolved from organic soil at different depths *in situ. Can. J. Soil Sci.*, 61:137–144.

Clark, J.S. and Nichol, W.E. 1966. The lime potential-percent base saturation relations of acid surface horizons of mineral and organic soils. *Can. J. Soil Sci.*, 46:281–285.

Cogger, C. and Duxbury, J.M. 1984. Factors affecting phosphorus losses from cultivated organic soils. *J. Environ. Qual.* 13:111–114.

CPVQ. 1996. *Crop Fertilization Guide* (in French). Agdex 540, 2nd ed., Conseil des Productions Végétales du Québec, Quebec, Canada.

Damman, A.W.H. 1988. Regulation of nitrogen and retention in *Sphagnum* bogs and other peatlands. *Oikos,* 51:291–305.

Daughtrey, Z.W., Gilliam, J.W., and Kamprath, E.J. 1973a. Phosphorus supply characteristics of acid organic soils as measured by desorption and mineralization. *Soil Sci.*, 115:18–24.

Daughtrey, Z.W., Gilliam, J.W., and Kamprath, E.J. 1973b. Soil test parameters for assessing plant-available P of acid organic soils. *Soil Sci.*, 115:438–446.

DeBusk, W.F. et al. 1994. Spatial distribution of soil nutrients in a northern Everglades marsh: Water conservation area 2A. *Soil Sci. Soc. Am. J.*, 58:543–552.

Diaz, O.A., Anderson, D.L., and Hanlon, E.A. 1993. Phosphorus mineralization from organic soils of the Everglades agricultural area. *Soil Sci.*, 156:178–185.

Droughty, J.L. 1930. The fixation of phosphate by a peat soil. *Soil Sci.*, 2:23–35.

Duxbury, J.M. and Peverly, J.H. 1978. Nitrogen and phosphorus losses from organic soils. *J. Environ. Qual.*, 7:566–570.

Duxbury, J.M. and Tate, R.L., III. 1981. The effect of soil depth and crop cover on enzymatic activities in Pahokee muck. *Soil Sci. Soc. Am. J.*, 45:322–328.

Fox, R.L. and Kamprath, E.J. 1971. Adsorption and leaching of P in acid organic soils and high organic matter sand. *Soil Sci. Soc. Am. Proc.*, 35:154–156.

Frissel, M.J. (Ed.). 1977. Cycling of mineral nutrients in agricultural ecosystems. *Agro-Ecosystems*, 4:1–354.

Godefroy, J. 1977. Physical and chemical analyses of peat soils (in French). *Fruits,* 32:647–664.

Gotkiewicz, J. 1977. Mineralization of nitrogen compounds in distinguished sites of the Wizna fen, in *Differentiation of Ecological Conditions of a Grassland Exosystem on the Wizna Fen as Influenced by Reclamation. Polish Ecological Studies,* Vol. 3(3). Okruszko, H. (Ed.), Polish Scientific Publ., Warsaw, Poland, 33–43.

Groffman, P.M. et al. 1996. Variation in microbial biomass and activity in four different wetland types. *Soil Sci. Soc. Am. J.*, 60:622–629.

Guthrie, T.F. and Duxbury, J.M. 1978. Nitrogen mineralization and denitrification in organic soils. *Soil Sci. Soc. Am. J.*, 42:908–912.

Hamilton, H.A. and Bernier, R. 1975. N-P-K fertilizer effects on yield, composition and residues of lettuce, celery, carrot and onion grown on an organic soil in Quebec. *Can. J. Plant Sci.*, 55:453–461.

Hamilton, H.A. and Bernier, R. 1973. Effects of lime on some chemical characteristics, nutrient availability, and crop response of a newly broken organic soil. *Can. J. Soil Sci.,* 53:1–8.

Harris, C.I. and Warren, G.F. 1962. Determination of phosphorus fixation capacity in organic soils. *Soil Sci. Soc. Am. Proc.,* 26:381–383.

Harris, C.I. et al. 1962. Water-level control in organic soil, as related to subsidence rate, crop yield, and response to nitrogen. *Soil Sci.,* 94:158–161.

Herlihy, M. 1972. Microbial and enzyme activity in peats. *Acta Hortic.,* 26:45–50.

Hortenstine, C.C. and Forbes, R.B. 1972. Concentrations of nitrogen, phosphorus, potassium, and total soluble salts in soil solution samples from fertilized and unfertilized organic soils. *J. Environ. Qual.,* 1:446–449.

Isirimah, N.O. and Keeney, D.R. 1973. Nitrogen transformations in aerobic and waterlogged organic soils. *Soil Sci.,* 115:123–129.

Ivarson, K.C. 1977. Changes in decomposition rate, microbial population and carbohydrate content of an acid peat bog after liming and reclamation. *Can. J. Soil Sci.,* 57:129–137.

Izuno, F.T. et al. 1991. Phosphorus concentrations in drainage water in the Everglades agricultural area. *J. Environ. Qual.,* 20:608–619.

Jenkinson, D.S. 1990. The turnover of organic carbon and nitrogen in soil. *Phil. Trans. R. Soc. Lond.,* B329:361–368.

Kaila, A. 1959a. Effect on a peat soil of application of superphosphate at various rates. *J. Sci. Agric. Soc. Finland,* 31:120–130.

Kaila, A. 1959b. Retention of phosphate by peat samples. *J. Sci. Agric. Soc. Finland,* 31:215–225.

Kaila, A. 1958a. Effect of superphosphate on the mobilization of nitrogen in a peat soil. *J. Sci. Agric. Soc. Finland,* 30:114–124.

Kaila, A. 1958b. Availability for plants of phosphorus in some virgin peat samples. *J. Sci. Agric. Soc. Finland,* 30:133–142.

Kaila, A. 1956. Phosphorus in various depths of some virgin peat lands. *J. Sci. Agric. Soc. Finland,* 28:90–104.

Kaila, A., Köylijärvi, J., and Kivinen, E. 1953. Influence of temperature upon the mobilization of nitrogen in peat. *J. Sci. Soc. Finland* 25:37–46.

Kaila, A. and Ryti, R. 1968. Effect of application of lime and fertilizers on cultivated peat soil. *J. Sci. Agric. Soc. Finland,* 40:133–141.

Kaila, A. and Soini, S. 1957. Influence of lime on the accumulation of mineral nitrogen in incubation experiments on peat soils. *J. Sci. Agric. Soc. Finland,* 29:229–237.

Kaila, A. and Missilä, H. 1956. Accumulation of fertilizer phosphorus in peat soils. *J. Sci. Agric. Soc. Finland,* 28:168–178.

Kaila, A., Soini, S. and Kivinen, E. 1954. Influence of lime and fertilizers upon the mineralization of peat nitrogen in incubation experiments. *J. Sci. Agric. Soc. Finland,* 26:79–95.

Kaunisto, S. and Aro, L. 1999. Forestry use of cut-away peatlands, in *Peatlands in Finland.* Vasander, H. (Ed.), Finnish Peatland Society, Helsinki. 130–137

Khiari, L., Parent, L.E., and Tremblay, N. 2001a. Selecting the high-yield sub-population for diagnosing nutrient imbalance in crops. *Agron. J.,* 93:802–808.

Khiari, L., Parent, L.E., and Tremblay, N. 2001b. Critical compositional nutrient indexes for sweet corn at early growth stage. *Agron. J.,* 93:809–814.

Khiari, L., Parent, L.E., and Tremblay, N. 2001c. The phosphorus compositional nutrient diagnosis range for potato. *Agron. J.,* 93:815–819.

Khiari, L. et al. 2000. An agri-environmental phosphorus saturation index for acid, light-textured, soils. *J. Environ. Qual.,* 29:1561–1567.

Kivekäs, J. 1958. The effect of drying and grinding of peat samples on the results of analyses. *J. Sci. Agric. Soc. Finland,* 30:1–9.

Kivekäs, J. and Kivinen, E. 1959. Observations on the mobilization of peat nitrogen in incubation experiments. *J. Sci. Agric. Soc. Finland,* 31:268–281.

Koch-Rose, M.S., Reddy, K.R. and Chanton, J.P. 1994. Factors controlling seasonal nutrient profiles in a subtropical peatland of the Florida Everglades. *J. Environ. Qual.,* 23:526–533.

Kong, K.T. and Dommergues, Y. 1970. Limitation of cellulolysis in organic soils, I. Respirometric studies (in French). *Rev. Écol. Biol. Sol,* 7:441–456.

Kong, K.T. and Dommergues, Y. 1972. Limitation of cellulolysis in organic soils, II. Enzyme studies (in French). *Rev. Écol. Biol. Sol,* 9:629–640.

Kong, K.T. and Dommergues, Y. 1973. Limitation of cellulolysis in organic soils, III. Competition between the cellulolytic and noncellulolytic microflora in organic soils (in French). *Rev. Écol. Biol. Sol,* 10:45–53.

Kuntze, H. and Eggelsmann, R. 1975. The influence of agriculture on eutrophication in lowland areas, in *Proc. Minsk Symp. on Hydrology of Marsh-Ridden Areas,* The UNESCO Press, IAHS, Paris. 479–486

Larsen, J.E., Warren, G.F., and Langston, R. 1959. Effect of iron, aluminium and humic acid on phosphorus fixation by organic soils. *Soil Sci. Soc. Am. Proc.,* 23:438–440.

Larsen, J.E., Langston, R. and Warren, G.F. 1958. Studies on the leaching of applied labeled phosphorus in organic soils. *Soil Sci. Soc. Am. Proc.,* 22:558–560.

Lawton, K. and Davis, J.F. 1956. The effect of liming on the utilization of osil and fertilizer phosphorus by several crops grown on acid organic soils. *Soil Sci. Soc. Am. Proc.,* 20:322–326.

Longbucco, P. and Rafferty, M.R. 1989. Delivery of nonpoint-source phosphorus from cultivated mucklands to Lake Ontario. *J. Environ. Qual.,* 18:157–163.

Lucas, R.E. and Davis, J.F. 1961. Relationships between pH values of organic soils and availabilities of 12 plant nutrients. *Soil Sci.,* 92:177–182.

MacLean, A.J. et al. 1964. Comparison of procedures for estimating exchange properties and availability of phosphorus and potassium in some eastern Canadian organic soils. *Can. J. Soil Sci.,* 44:66–75.

Macmillan, K.A. 1976. Biological oxidation of a humic Mesisol under laboratory conditions. *Can. J. Soil Sci.,* 56:51–53.

Mangler, J.E. and Tate, R.L., III. 1982. Source and role of peroxidase in soil organic matter oxidation in Pahokee muck. *Soil Sci.,* 134:226–232.

Martikainen, P.J., et al. 1995. Change in fluxes of carbon dioxide, methane and nitrous oxide due to forest drainage of mire sites of different trophy. *Plant Soil,* 168–169:571–577.

Mathur, S.P. and Sanderson, R.B. 1980. The partial inactivation of degradative soil enzymes by residual fertilizer copper in histosols. *Soil Sci. Soc. Am. J.,* 4(4):750–755.

McGill, W.B. and Cole, C.V. 1981. Comparative aspects of cycling of organic C, N, S and P through soil organic matter. *Geoderma,* 26:267–286.

Mehlich, A. 1984. Mehlich-3 soil test extractant: a modification of Mehlich-2 extractant. *Comm. Soil Sci. Plant Anal.,* 15:1409–1416.

Miller, M.H. 1979. Contribution of nitrogen and phosphorus to subsurface drainage water from intensively cropped mineral and organic soils in Ontario. *J. Environ. Qual.,* 8:42–48.

Minotti, P.L. and Stone, K.W. 1988. Consequences of not fertilizing onions on organic soils with high soil test values. *Comm. Soil Sci. Plant Anal.,* 19:1887–1906.

Murayama, S., Asakawa, Y., and Ohno, Y. 1990. Chemical properties of subsurface peats and their decomposition kinetics under field conditions. *Soil Sci. Plant Nutr.,* 36:129–140.

Neller, J.R. 1944. Oxidation loss of lownoor peat in fields with different water tables. *Soil Sci.,* 58:195–204.

Nesterenko, I.M. 1976. Subsidence and wearing out of peat soils as a result of reclamation and agricultural utilization of marshlands. *Proc. 5th Int. Peat Congr.,* 1:218–232.

Nicholls, K.H. and MacCrimmon, H.R. 1974. Nutrients in subsurface and runoff waters of the Holland Marsh, Ontario. *J. Environ. Qual.,* 3:31–35.

Okruszko, H., Wilcox, G.E., and Warren, G.F. 1962. Evaluation of laboratory tests on organic soils for prediction of phosphorus availability. *Soil Sci. Soc. Am. Proc.,* 26:71–74.

Parent, L.E., Millette J.A., and Mehuys, G.R. 1982. Subsidence and erosion of a histosol. *Soil Sci. Soc. Am. J.,* 46:404–408.

Parent, L.E. and MacKenzie, A.F. 1985. Rate of pyrophosphate hydrolysis in organic soils. *Can. J. Soil Sci.,* 65:497–506.

Parent, L.E., Viau, A., and Anctil, F. 2000. Nitrogen and phosphorus fractions as indicators of organic soil quality. *Suo (Finland),* 51:71–81.

Parent, L.E. et al. 1997. Row-centred log ratios as nutrient indexes for saturated extracts of organic soils. *Can. J. Soil Sci.,* 77:571–578.

Parent, L.E. et al. 1992. The P status of cultivated organic soils in Quebec. *Proc. 9th Int. Peat Congr.* (Uppsala, Sweden), 2:400–410.

Parent, L.E., Grenon, L., and Buteau, P. 1991. Effect of many decades of intensive agricultural utilization on properties of organic soils in south-western Quebec (in French). *Proc. Quebec Symp. '89, "Peat and peatlands. Diversification and innovation.",* II:92–98.

Parton, W.J., Stewart, J.W.B., and Cole, C.V. 1988. Dynamics of C, N, P, and S in grassland soils: a model. *Biogechem,* 5:109–131.

Pons, L.J. 1960. Soil genesis and classification of reclaimed peat soils in connection with initial soil formation. *Proc. 7th Int. Congr. Soil Sci.,* 28:205–211.

Porter, P.S. and Sanchez, C.A. 1992. The effect of soil properties on phosphorus sorption by Everglades histosols. *Soil Sci.* 154:387–398.

Puustjärvi, V. 1970. Mobilization of nitrogen in peat culture. *Peat Plant News,* 3/1970:35–42.

Racz, G.J. 1979. Release of P in organic soils under aerobic and anaerobic conditions. *Can. J. Soil Sci.,* 59:337–339.

Reddy, K.R. 1983. Soluble phosphorus release from organic soils. *Agric. Ecosyst. Environ.,* 9:373–382.

Richardson, C.J. and Vaithiyanathan, P. 1995. Phosphorus sorption characteristics of Everglades soils along a eutrophication gradient. *Soil Sci. Soc. Am. J.,* 59:1782–1788.

Riekels, J. 1977. Nitrogen-water relationship of onions grown on organic soil. *J. Amer. Soc. Hort. Sci.,* 102:139–142.

Robinson, J.A. 1985. Determining microbial kinetic parameters using nonlinear regression analysis. Advantages and limitations in microbial ecology. *Adv. Microb. Ecol.,* 8:61–114.

Ross, S.M. 1987. *Introduction to Probability and statistics for Engineers and Scientists.* John Wiley & Sons, New York.

Sahrawat, K.L. 1983. Organic nitrogen in potassium chloride extracts of organic soils in the Philippines. *Geoderma,* 29:77–80.

Salonen, M., Koskela, I., and Kähäri, J. 1973. The dependence of the phosphorus uptake of plants on the properties of the soil. *Annales Agriculturae Fenniae,* 12:161–171.

SAS Institute, Inc. 1989. *Property Software, Release 6.08.* SAS Institute, Inc., Cary, NC.

Sasseville, L. 1991. Influence of liming on physico-chemical and biochemical properties and the phosphorus status in the surface horizon of an acid organic soil (in French). M. Sc. Thesis, Laval University, Quebec.

Scheffer, B. 1976. Nitrogen transformations in fen soils (in German). *Landwirtschaftliche Forschung,* 33:20–28, J.D. Sauerländer's Verlag, Frankfurt-am-Main, Germany.

Scheffer, B. and Kuntze, H. 1989. Phosphate leaching from high moor soils. *Int. Peat J.,* 3:107–115.

Schothorst, S. 1977. Subsidence of low moor peat soils in the western Netherlands. *Geoderma,* 17:265–291.

Simoneau, M. 1996. *Water Quality of the Châteauguay River Basin, 1979 à 1994* (in French). Direction des Écosystèmes aquatiques, Ministère de l'Environnement et de la Faune du Québec, Québec.

Sowden, F.J., Morita, H., and Levesque, M. 1978. Organic nitrogen distribution in selected peats and peat fractions. *Can. J. Soil Sci.,* 58:237–249.

Stevenson, F.J. 1986. *Cycles of Soils. C, N, P, S, Micronutrients.* John Wiley & Sons, New York.

Tate III, R.L. 1987. *Soil Organic Matter. Biological and Ecological Effects.* John Wiley & Sons, New York.

Tate, R.L., III, 1984. Function of protease and phosphatase activities in subsidence of Pahokee muck. *Soil Sci.,* 138:271–278.

Tate, R.L., III. 1980a. Variation in heterotrophic and autotrophic nitrifier populations in relation to nitrification in organic soils. *Appl. Environ. Microbiol.,* 40(1):75–79.

Tate, R.L., III. 1980b. Microbial oxidation of organic matter of histosols. *Adv. Microb. Ecol.* 4:169–201.

Tate, R.L., III. 1980c. Effect of several environmental parameters on carbon metabolism in histosols. *Microb. Ecol.,* 5:329–336.

Tate, R.L., III. 1979. Microbial activity in organic soils as affected by soil depth and crop. *Soil Sci.,* 37:1085–1090.

Tate, R.L., III. 1976. Nitrification in Everglades organic soils: a potential role in soil subsidence. *Proc. Anaheim Symp., Int. Assoc. Hydrol. Sci.,* Publ. 121:657–663.

Terry, R.E. 1980. Nitrogen mineralization in Florida organic soils. *Soil Sci. Soc. Am. J.,* 44:747–750.

Terry, R.E., and Tate, R.L., III. 1980a. The effect of nitrate on nitrous reduction in organic soils and sediments. *Soil Sci. Soc. Am. J.,* 44:744–746.

Terry, R.E., and Tate, R.L., III. 1980b. Denitrification as a pathway for nitrate removal from organic soils. *Soil Sci.,* 129:162–166.

Tremblay, N. and Parent, L.E. 1989. Residual effect N, P and K fertilizers on yield of carrots and onions in organic soils (in French). *Naturaliste Can.,* 116:131–136.

Turk, L.M. 1943. Effect of certain mineral elements on some microbiological activities in muck soils. *Soil Sci.,* 47:425–445.

Van der Zee, S.E.A.T.M., Fokkink, L.G.J., and Van Riemsduk, W.H. 1987. A new technique for assessment of reversibly adsorbed phosphate. *Soil Sci. Soc Am. J.,* 51:599–604.

Van Heuveln, B., Jongerius, A., and Pons, L.J. 1960. Soil formation in organic soils. *Proc. 7th Int. Congr. Soil Sci.,* 27:196–205.

Van Lierop, W. 1981. Conversion of organic soil pH values measured in water, 0.01 *M* $CaCl_2$ or 1 N KCl. *Can. J. Soil Sci.,* 61:577–579.

Van Lierop, W. and MacKenzie, A.F. 1977. Soil pH measurement and its application to organic soils. *Can. J. Soil Sci.,* 57:55–64.

Verhoeven, J.T.A., Kooijman, A.M., and Van Wirdum, G. 1988. Mineralization of N and P along a trophic gradient in a freshwater mire. *Biogeochem.,* 6:31–43.

Volk, B.G. 1973. Everglades histosol subsidence, 1. CO_2 evolution as affected by soil type, temperature, and moisture. *Soil Crop Sci. Soc. Florida Proc.,* 32:132–135.

Volk, B.G. and Sartain, J.B. 1976. Elemental concentrations of drainage water from Everglades organic soils as affected by cropping systems. *Soil Crop Sci. Soc. Florida Proc.,* 35:177–183.

Wagar, B.I., Stewart, J.W.B., and Moir, J.O. 1986. Changes with time in the form and availability of residual fertilizer phosphorus on chernozemic soils. *Can. J. Soil Sci.,* 66:105–119.

Waksman, S.A. and Purvis, E.R. 1932a. The influence of moisture upon the rapidity of decomposition of lowmoor peat. *Soil Sci.,* 34:323–336.

Waksman, S.A. and Purvis, E.R. 1932b. The microbiological population of peat. *Soil Sci.,* 34:95–100.

Walbridge, M.R. 1991. Phosphorus availability in acid organic soils of the lower North Carolina coastal plain. *Ecology,* 72:2083–2100.

Walker, T.W. and Adams, A.F.R. 1958. Studies on soil organic matter, 1. Influence of phosphorus content of parent materials on accumulations of carbon, nitrogen, sulfur, and organic phosphorus in grassland soils. *Soil Sci.,* 85:307–318.

Warncke, D.D. 1990. Testing artificial growth media and interpreting the results, in *Soil Testing and Plant Analysis.* 3rd ed. Westermann, R.L. (Ed.), Soil Sc. Soc. Am. Book Ser. 3, Madison, WI, 337–357.

Waughman, G.J. 1980. Chemical aspects of the ecology of some South German peatlands. *J. Ecol.,* 68:1025–1046.

Williams, B.L. 1974. Effect of water-table level on nitrogen mineralization in peat. *Forestry,* 47:195–202.

Williams, B.L. and Sparling, G.P. 1984. Extractable N and P in relation to microbial biomass in UK acid organic soils. *Plant Soil,* 76:139–148.

Williams, B.L. and Silcock, D.J. 2001. Does nitrogen addition to raised bogs influence peat phosphorus pools? *Biogeochem.,* 53:307–321.

Williams, B.L. and Silcock, D.J. 1997. Nutrient and microbial changes in the peat profile beneath *Sphagnum magellanicum* in response to additions of ammonium nitrate. *J. Appl. Ecol.,* 34:961–970.

Zimenko, T.G. and Revinskaya, L.S. 1972. Activity of microorganisms in peat-bog soils. *Mikrobiol.,* 41:891–895.

CHAPTER 7

Retention of Copper in Cu-Enriched Organic Soils

Antoine Karam, Caroline Côté, and Léon E. Parent

CONTENTS

ABSTRACT

Copper may accumulate in organic soils in the range of 2 to 60,000 mg kg^{-1}, naturally or as a result of fertilizer or biocide applications. The authors conducted a study on Cu sorption and extraction using 28 moorsh materials varying in quality attributes. The extraction sequence included water soluble and exchangeable Cu. Sorption was described by the Langmuir equation with maximum sorption capacity (X_m) in the range of 24 to 55 g Cu kg^{-1}. The X_m was quartically related to the sum of exchangeable basic cations (SEBC) ($R^2 = 0.97$). Three sorption patterns were found: X_m was constant for SEBC values below 45 $cmol_c$ $kg,^{-1}$ then increased in proportion of SEBC up to 85 $cmol_c$ $kg,^{-1}$ and finally increased at a lower rate for higher SEBC values. The H_2O- and KNO_3-extractable Cu from added Cu at assumed toxic level (3000 mg Cu kg^{-1}) was cubically related to SEBC and pH; it was highest below a SEBC value of 45 $cmol_c$ kg^{-1} or a pH (0.01 M $CaCl_2$) value of 4.2, then declined to reach a plateau. The Cu sorption and desorption capacities in organic soils can be assessed from easily determined properties such as SEBC and pH.

I. INTRODUCTION

The Cu content is generally low in organic soils (Lévesque and Mathur, 1983a; Mengel and Rehm, 2000) compared with mineral soils (Jasmin and Hamilton, 1980). In Canada, Cu content varied from 1.9 mg kg^{-1} in a Newfoundland bog (Mathur and Rayment, 1977) to 60,000 mg kg^{-1} in a cupriferrous New Brunswick bog (Boyle, 1977); however, normal Cu content is in the range of 8.3 to 537.5 mg total Cu kg^{-1} in Canadian moorsh soils (Lévesque and Mathur, 1986; Mathur et al., 1989).

The strong ability of humic substances (HS) to form stable complexes with Cu is a major cause of Cu deficiency in soils (Matsuda and Ikuta, 1969; Mortvedt, 2000). Organic soils containing less than 20–30 mg total Cu kg^{-1} in the moorsh layers are considered deficient (Lucas, 1982). The recommended Cu application rates in organic soils range between 10 and 20 kg Cu ha^{-1} every 3 years (CPVQ, 1996). At such rates, Cu is harmless to the environment (Hamilton, 1979; Mathur et al., 1979a; Preston et al., 1981).

The Cu may accumulate to levels exceeding agronomic requirements either naturally or through human activities. The Cu enrichment in peats is due in part to the formation of stable complexes with organic macromolecules (Leeper, 1978; Shotyk et al., 1992). The HS can release Cu in amounts suitable for plant growth (Donahue et al., 1983; Tan, 1998). The $Na_4P_2O_7$-extractable Cu, whereby humic acids (HA) and fulvic acids (FA) are also extracted, is thus considered to be the most available form to plants (Viets, 1962); however, Cu linked to HA and humins is considered to be less available to the plants than Cu linked to the lower molecular weight FA (Preston et al., 1981; Schnitzer and Khan, 1972; Szalay et al., 1975). Brennan et al. (1980) found that availability of freshly applied Cu to wheat decreased by 70% with incubation time up to 120 days. Brennan et al. (1983) also found that fresh wheat straw decreased Cu availability when applied at rates of 2.5 to 10 g per 100 g in a Lancelin soil containing 0.8% organic matter (OM).

The aim of this chapter is to examine the effect of soil properties on Cu sorption and desorption in Cu-enriched moorsh soils.

II. CU MOBILITY AND TOXICITY

In organic and acid mineral soils, soil organic matter (SOM) is the dominant Cu sorbent (Stevenson, 1982). Because peats are known to sequestrate Cu (Boyle, 1977), to sorb high amounts of applied Cu (Parent and Perron, 1983), and to form stable complexes with Cu (Basu et al., 1964; Bunzl et al., 1976; Schnitzer, 1978), low to moderate Cu additions are unlikely to contribute to the pollution of groundwater (Hamilton, 1979; Mathur et al., 1979a) or to initiate Cu leaching (Preston et al., 1981). The formation of metal-organic complexes must influence the concentration and mobility of Cu^{2+} in soils (Cavallaro and McBride, 1978). At high Cu rates and in presence of high amounts of FA, Cu is mainly sequestered as soluble organic complexes (McBride and Blasiak, 1979; McLaren et al., 1981). The humus immobilizes a high proportion of the Cu applied at a low rate. Intensive decomposition of humus or oxidation of moorsh soils must contribute to the release of Cu from humates in a form more available to plants; however, Cu is generally considered as relatively immobile in organic soils.

Phytotoxicity of soil Cu is controlled by sorption and desorption reactions as related to pH, cation exchange capacity (CEC), SOM content, and the soil capacity to supply P, Ca, and Fe to plants (Leeper, 1978; Mathur and Lévesque, 1983). Sorbed Cu is partially reversible (Kadlec and Rathbun, 1983), therefore, Cu may become toxic above a threshold concentration. The threshold of Cu phytoxicity in organic soils can be predicted to some extent by CEC. Lévesque and Mathur (1984) concluded that the threshold of soil-Cu toxicity in vegetable crops was about 5% of CEC or 16 mg total Cu kg^{-1} for each $cmol_c$ kg^{-1} of CEC as determined by the neutral ammonium acetate method.

Bear (1957) found that applications of as much as 11,200 kg Cu ha^{-1} or 28,000 mg Cu kg^{-1} to organic soil materials containing low amounts of plant-available Cu did not retard plant growth. Plants not responding strongly to Cu can be grown in moorsh soils containing up to 1063 mg kg^{-1} of Cu without adverse effects on yield (Mathur and Lévesque, 1983). An experiment involving the application of Cu to moorsh soils in amounts that result in EDTA-Cu levels more than 1148 times the plant requirements did not increase Cu concentration in oat grain or straw (Mathur et al., 1979a). Lévesque and Mathur (1983a) concluded that the enrichment of moorsh soils up to 100 mg Cu kg^{-1} are not phytotoxic.

Copper mitigates subsidence through its ability to inactivate degradative soil enzymes taking part in SOM mineralization (Bowen, 1966; Mathur and Rayment, 1977; Mathur and Sanderson, 1978; Mathur et al., 1979b; Mathur et al., 1980; Mathur, 1983). Levels of 100, 200, 300, and 400 mg total Cu kg^{-1} in organic soils with bulk densities of 0.1, 0.2, 0.3, and 0.4 g cm^3, respectively, must be maintained in order to reduce the subsidence rate by 50% (Mathur et al., 1979b; Mathur, 1982a, b; Lévesque and Mathur, 1984). A rate of 100 kg Cu ha^{-1} during the first few years of cultivation is effective in mitigating subsidence (Mathur et al., 1979b; Preston et al., 1981). In comparison, up to 15 kg Cu ha^{-1} are normally applied yearly to newly reclaimed organic soils during the first 3 years of cultivation, and then 5 kg Cu ha^{-1} every second or third year (Lévesque and Mathur, 1984). Lévesque and Mathur (1983b, 1984) reported that Cu addition at three times the rate for mitigating subsidence by about 50% would not adversely affect the growth or nutrition of crops grown in this soil.

III. CU SORPTION

A. Theory

The Cu content in plants is controlled mainly by Cu concentration in the soil solution as determined by sorption reactions (McLaren and Crawford, 1973). Sorption of Cu is influenced by many soil properties such as HS, clay, carbonate, as well as oxides of Al, Fe, and Mn, pH, CEC, exchangeable cations, mineralogy, ionic strength, and soil solution composition (Kishk and Hassan, 1973; Harter, 1979; Dhillon et al., 1981; Duquette and Hendershot, 1990; Basta and Tabatabai, 1992a). The ability of HA and FA to remove trace metals from solution is well documented (Basu et al., 1964; Ellis and Knezek, 1972; Rachid, 1974; Christensen et al., 1998; Ravat et al., 2000). Sorption of Cu by organic soils occurs at a high rate, depending on the initial concentration of Cu in solution (Sapek, 1976; Sapek and Zebrowski, 1976).

Metal binding sites on HS are heterogeneous (Schnitzer, 1969; Petruzzelli et al., 1981; Murray and Linder, 1983; Christensen et al., 1998). The HA in peat (Szalay and Szilágyi, 1968) is stable and highly reactive (Senesi et al., 1989). Goodman and Cheshire (1976) as well as Abdul-Halim et al. (1981) suggested that small quantities of Cu^{2+} are tightly bound to HA through a porphyrin-type linkage. Interactions of Cu with HS involve outer sphere complexation (electrostatic attraction), ion exchange, inner sphere complexation, precipitation, and dissolution as a function of acidic functional groups in HS, pH, and ionic strength (McBride, 1994, Kabata-Pendias, 2001). Because Cu can form inner-sphere complexes with organic ligands (Sposito, 1984), more Cu must remain in soil solution as competition with H^+ ions increases. Manganese, Fe, and Al oxides can sorb Cu^{2+} more strongly than most divalent metals (McBride, 2000). The Mn oxides show high selectivity for Cu^{2+} (McKenzie, 1980); however, chelated Mn in moorsh soils (Lévesque and Mathur, 1983b) can be easily displaced by Cu^{2+}. The Fe and Mn oxides and hydroxides adsorb trace metals due to their high surface areas coupled with the ability of Cu^{2+} to replace Fe^{2+} in some Fe-oxides (Taylor, 1965; Tessier et al., 1979; Hickey and Kittrick, 1984).

B. Experimental Setup

The authors conducted two laboratory experiments on Cu sorption and desorption using 28 moorsh soil materials (0–15 cm) from southwestern Quebec, Canada, and showing a wide range of chemical properties. Soil samples were air-dried, sieved to <2 mm, and stored at room temperature until use. Soil properties were determined by standard methods (McKeague, 1978; Karam, 1993), and included pH in $CaCl_2$ 0.01 M; SOM by combustion (550°C), ammonium acetate-extractable Ca, Mg, and K; acid–ammonium–oxalate-extractable Fe, Al, and Mn; and sodium pyrophosphate-extractable Fe, Al, and Mn. Acid ammonium oxalate extracts amorphous Fe, Al, and Mn oxides and metals complexed to SOM, whereas sodium pyrophosphate extracts Fe, Al, and Mn associated with SOM (McKeague, 1978). The CEC at pH 7.0 was calculated as the SEBC plus exchangeable acidity. Total Cu was determined after acid digestion (Page et al., 1982). Metal (Cu, Fe, Al, and Mn) concentrations were determined by atomic absorption spectrophotometry.

Chemical properties are presented in Table 7.1. The CEC varied between 57 and 187 $cmol_c$ kg^{-1}, averaging 148 $cmol_c$ kg^{-1}, which is a value between those obtained by Lévesque and Mathur (1986) for other Quebec moorsh soils, and by MacLean et al. (1964) for other eastern Canadian organic soils. Exchangeable acidity varied from 12 to 141 $cmol_c$ kg^{-1}, averaging 52% of CEC. A linear relationship existed between OM content and CEC (r = 0.868, $P < 0.001$). Total Cu content ranged from 9 to 79 mg kg^{-1}, averaging 44 mg kg^{-1}, lower than the mean value of 145 (8 to 538) obtained by Mathur and Lévesque (1989). The pH averaged 5.4 in the range of 4.1 to 6.7.

The Cu additions varied between 3000 and 60,000 mg Cu kg^{-1}, the maximum value for Cu accumulation in organic soils (Boyle, 1977). The 3000 mg Cu kg^{-1} treatment exceeded slightly the toxic level of 16 mg total Cu kg^{-1} per $cmol_c$ kg^{-1} of CEC (Levesque and Mathur, 1984) for the highest CEC value among our moorsh soils (187 $cmol_c$ kg^{-1} times 16 is 2992 mg Cu kg^{-1}). Sorption of Cu was conducted as follows: one-gram sample was weighed into each of six 50-mL polyethylene centrifuge tubes and 30 mL of a 0.01 M $CaCl_2.2H_2O$ solution containing 0, 100, 500, 1000, 1500, or 2000 mg L^{-1} of Cu as $CuSO_4.2H_2O$, thus providing 0; 3000; 15,000; 30,000; 45,000; or 60,000 mg Cu kg^{-1} soil, respectively. Soil suspensions were allowed to equilibrate for 48 h at room temperature with occasional shaking, then they were centrifuged and decanted. The supernatant solution was analyzed for Cu by atomic absorption spectrophotometry. The amount of Cu sorbed was computed as the difference between initial concentration added and that remaining in the supernatant solution. Sorption maximum capacity (X_m) was computed using the linearized version of the Langmuir equation (Bohn et al., 2001) as follows:

$$C/X = 1/(kX_m) + CX_m \qquad (7.1)$$

where C is Cu concentration in the equilibrium solution (mg L^{-1}), X is the amount of Cu sorbed (mg kg^{-1}), X_m is maximum sorption capacity (mg Cu kg^{-1} soil), and k is a constant thought to be related to the bonding energy (L mg^{-1} sorbed Cu). Sorption isotherms showed a goodness-of-fit (r^2) of 0.993 or more (SAS Institute, 1990). The X_m is of agronomic and environmental importance as a measure of the soil capacity to retain Cu. The k value is difficult to interpret in multisite systems. Parameters k and X_m may not have any particular chemical meaning when a reaction mechanism other than adsorption occurs (Veith and Sposito, 1977).

C. Results and Discussion

The X_m varied from 24,326 to 55,157 mg kg^{-1} (Table 7.1), in agreement with those obtained by Goodman and Cheshire (1973), and Parent and Perron (1983). With the exception of few samples, mean values for X_m were lower the more acidic the soil. Assuming that Cu was sorbed as Cu^{2+}, X_m varied between 77 and 174 $cmol_c$ kg^{-1}, averaging 85% of CEC; 24 soils gave X_m values in the range of 46.1 to 98.3% of CEC.

Table 7.1 Chemical Properties of 28 Organic Soil Materials from Southwestern Quebec

Sample	SOM (%)	pH	Ca^{2+} exch (cmol$_c$ kg⁻¹)	Mg^{2+} exch	K^+ exch	SEBC	H^+ exch	CEC	Fe_{ox} (mg kg⁻¹)	Fe_{pyr}	Al_{ox}	Al_{pyr}	Mn_{ox}	Mn_{pyr}	Cu	X_m	X_m (cmol$_c$ kg⁻¹)
1	77.7	5.74	53.3	6.7	1.35	61.4	71.1	133	4966	3255	4544	3440	176	137	63	37,891	119
2	79.6	4.80	52.1	8.6	0.53	61.2	98.7	160	9946	7042	4594	3460	152	143	56	35,122	111
3	33.1	6.58	39.1	5.2	0.61	44.9	12.5	57	5510	3033	1677	1215	220	163	35	25,491	80
4	89.6	5.55	56.4	9.7	1.41	67.5	78.4	146	1613	333	1191	822	59	47	35	39,093	123
5	91.1	4.09	27.5	4.9	1.80	34.2	131.9	166	2241	1228	1486	1362	43	62	9	24,326	77
6	91.1	4.96	56.4	12.6	0.96	70.0	95.8	166	3962	1713	1468	1135	262	232	10	41,862	132
7	92.2	4.72	46.7	13.1	1.38	61.2	115.1	176	1190	133	715	485	50	43	24	37,717	119
8	83.1	6.69	90.8	7.8	0.36	99.0	66.4	165	8802	3763	3322	2025	704	473	34	50,648	160
9	77.0	5.51	61.9	10.7	1.38	74.0	62.3	136	3378	1587	1938	1423	122	110	65	43,286	136
10	92.4	4.07	29.3	7.2	2.06	38.6	140.6	179	1118	515	1194	1042	43	58	17	28,423	90
11	84.8	5.15	63.2	17.5	0.55	81.3	73.9	155	2466	795	1269	898	106	88	31	48,347	152
12	83.8	6.02	82.8	12.0	1.12	95.9	65.1	161	5598	3658	1859	1432	309	138	57	50,100	158
13	83.7	5.43	56.3	17.5	0.41	74.2	77.9	152	4370	2795	2419	1845	207	160	23	44,595	140
14	84.7	5.90	62.8	9.6	0.63	73.0	72.7	146	7878	5633	3014	2487	408	342	72	43,689	138
15	86.9	5.42	55.9	8.8	1.25	66.0	75.9	142	2233	837	1841	1245	93	80	77	37,896	119
16	91.5	4.88	45.3	6.4	1.48	53.2	91.1	144	1402	472	1209	957	137	108	42	31,080	98
17	77.3	6.07	61.8	5.6	2.22	69.6	50.5	120	3010	1303	1427	1075	74	55	70	44,454	140
18	86.1	5.90	60.2	7.3	1.88	69.4	70.0	139	2199	847	1203	957	58	53	47	42,033	132
19	89.0	5.68	60.8	16.1	1.31	78.2	71.4	150	4222	2347	1814	1483	284	252	79	45,623	144
20	92.7	4.66	39.4	10.4	1.44	51.2	102.8	154	2197	842	978	755	35	30	31	30,815	97
21	87.3	5.65	64.6	8.0	1.12	73.7	72.2	146	8910	6477	3363	2908	424	383	64	43,316	136
22	86.9	5.37	74.9	13.3	1.66	89.9	76.0	166	4194	1862	1153	820	540	442	36	48,940	154
23	79.8	5.85	91.0	16.7	0.74	108.4	78.4	187	3001	1435	957	725	145	123	48	55,157	174
24	92.2	4.42	41.5	11.4	0.82	53.7	115.8	170	2742	1178	774	607	78	77	29	34,563	109
25	90.7	4.64	48.9	10.0	1.65	60.6	106.0	167	1948	840	1478	1417	82	73	28	38,959	123
26	45.2	6.36	42.8	13.3	0.77	56.9	31.1	88	6266	3425	3419	1953	249	188	52	31,454	99
27	59.0	5.57	45.8	11.6	0.66	58.1	57.7	116	6890	4780	3480	2290	235	210	71	34,014	107
28	86.3	5.39	66.5	14.7	0.89	82.1	74.1	156	3000	1587	2582	1953	119	90	39	48,819	154

Note: SOM = soil organic matter content; pH = pH in 0.01M CaCl$_2$; exch = exchangeable; SEBC = sum of exchangeable cations; CEC = cation exchange capacity; ox = oxalate; pyr = pyrophosphate; Cu = total Cu; X_m = Langmuir sorption maximum.

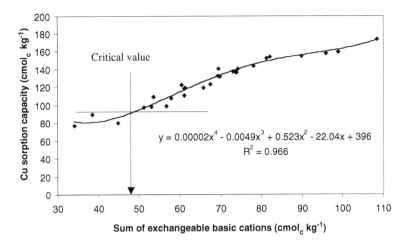

Figure 7.1   Relationship between the sum of exchangeable cations and the Langmuir Cu sorption capacity in cultivated organic soils.

Average pH of equilibrium solutions decreased from 6.46 (control) to 4.06 (highest Cu rate), which indicated competition between Cu^{2+} and H^+, for organic sorption sites, particularly in the higher concentration range of Cu (Beckwith, 1959; Schnitzer and Skinner, 1963; Khan, 1969), or a change in Cu hydrolysis state with an increasing Cu application rate (Jarvis, 1981; Basta and Tabatabai, 1992b).

Soil pH and related soil parameters, such as SEBC and CEC, can control the behavior of Cu in organic soils receiving appreciable amounts of Cu (Ravat et al., 2000). The X_m values were highly correlated with SEBC, and exchangeable calcium (Exch-Ca), but weakly correlated to pH, CEC, oxalate-extractable Mn (Mn_{ox}), and pyrophosphate-extractable Mn (Mn_{pyr}). A high correlation with SEBC ($r = 0.965$, $P < 0.001$) was also found by Harter (1979) and Basta and Tabatatai (1992a, 1992b). Indeed, Cu^{2+} can replace Ca^{2+} and Mg^{2+} on exchange sites (Harter, 1992). According to Alberts and Giesy (1983), Cu competes effectively with Ca for binding sites due to its higher stability constant with organics compared to Ca. The quartic relationship between SEBC and X_m (Figure 7.1) shows three sorption patterns: a plateau of constant sorption capacity for SEBC < 45 $cmol_c$ kg^{-1}, a trend of a high rate of increase in Cu sorption capacity between 45 and 85 $cmol_c$ kg^{-1}, followed by a trend of a smaller rate of Cu sorption capacity with SEBC > 85 $cmol_c$ kg^{-1}. As SOM contents come closer to 30% (soil 3) or SEBC values drop to less than 40 $cmol_c$ kg^{-1} (soils 5 and 10), Cu sorption capacity decreases markedly (Table 7.1).

IV. CU DESORPTION

A. Theory

Water soluble and exchangeable forms of Cu^{2+} are important sources of Cu for crop production. Briefly, those Cu forms are sequentially extracted using distilled water (H_2O) for 2 h, followed by 0.5 M KNO_3 for 16 h (Sposito et al., 1982). The water-soluble (H_2O) plus exchangeable (KNO_3) Cu content is widely regarded as a satisfactory measure of the ability of a soil to supply cationic micronutrients for plant growth (Lévesque and Mathur, 1988), therefore, high loads of Cu may produce toxic amounts of available Cu and perhaps also leachable Cu. As a result, desorption of water-soluble and exchangeable Cu is also crucial in environmental chemistry (Boyle, 1977). Schnitzer and Khan (1972) emphasized the importance of initial soil pH on availability and mobility of $Cu_{H2O+KNO3}$. Verloo et al. (1973) found that desorption and mobilization of soil Cu became significant as equilibrium pH fell toward 3.0.

B. Experimental Setup

The addition of 3000 mg Cu kg^{-1} increased CEC saturation from 0.10 ± 0.06% in the control to 6.9 ± 2.3% in Cu-treated samples in average, thus close to the 5% phytotoxicity threshold proposed by Lévesque and Mathur (1984). The water-soluble and exchangeable Cu (Sposito et al., 1982) was examined in soils treated with the 3000 mg Cu kg^{-1} application rate, which was slightly above the toxic level.

C. Results and Discussion

As shown in Figure 7.2 for moorsh soil materials containing more than 45% SOM, $Cu_{H2O+KNO3}$ decreased cubically with SEBC. As SEBC decreased, Cu competed more with protons. A critical value for $Cu_{H2O+KNO3}$ was found graphically at a SEBC of 45 $cmol_c$ kg^{-1}, in keeping with the lower critical value for Cu sorption

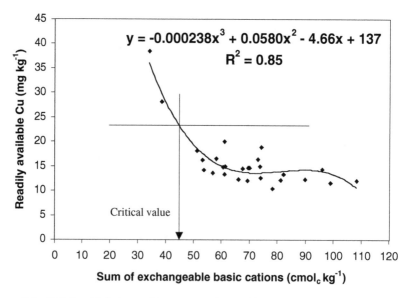

Figure 7.2   Relationship between the sum of exchangeable cations and readily available (sum of the H_2O and KNO_3 fractions) Cu from added Cu at toxic level (3000 kg mg^{-1}).

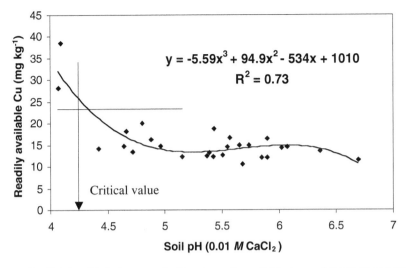

Figure 7.3   Relationship between soil pH and readily available (sum of the H_2O and KNO_3 0.5 M fractions) Cu from added Cu at toxic level (3000 mg kg).

(Figure 7.1). Thereafter, $Cu_{H2O+KNO3}$ decreased. Negative relationships between $Cu_{H2O+KNO3}$ and soil pH ($r = -0.80$, $P < 0.001$) or soil parameters related to pH, such as exchangeable Ca ($r = -0.78$, $P < 0.001$), SEBC ($r = -0.75$, $P < 0.001$), as well as the positive relationship between $Cu_{H2O+KNO3}$ and exchangeable acidity ($r = 0.71$, $P < 0.001$), provided further evidence that acid conditions exerted a dominant influence on the desorption of loosely bound Cu in Cu-enriched moorsh soils.

In fact, pH was by far the most important parameter, accounting for almost 63.5% of the variation in $Cu_{H2O+KNO3}$ values. Leeper (1978) emphasized that Cu is retained more weakly when the soil pH is lower. According to Tyler and McBride (1982), the mobility of metals in soils is determined by several factors, including the soil pH; however, metals (Cd, Cu, Ni, and Zn) move less readily through an acid organic soil (typic medisaprist) compared with mineral soils, presumably because of its high SOM content, sum of SEBC per unit volume, and CEC. Despite this, even a multiple linear regression incorporating pH, exchangeable acidity, Mn_{ox} (ox = oxalate), and Mn_{pyr} (pyr = pyrophosphate) accounted for only 18.2% more to the variation in $Cu_{H2O+KNO3}$ compared to pH alone.

The critical pH (0.01 M $CaCl_2$) for decreasing availability of Cu added to organic soils was 4.2 (Figure 7.3). Thus, pH (0.01 M $CaCl_2$) above 4.2 is an indicator of decreased Cu mobility in organic soils. Moorsh soil management is conducted at pH values higher than 4.2, therefore, Cu is not likely to cause toxicity or pollution problems under the present system of moorsh management.

V. CONCLUSION

In organic soils, three sorption patterns were defined: constant X_m for SEBC values below 45 $cmol_c$ kg^{-1}, increasing X_m in proportion of SEBC up to 85 $cmol_c$

kg^{-1}, and increasing X_m at a lower rate for higher SEBC values. Conversely, readily available Cu was highest below a SEBC value of 45 $cmol_c$ kg^{-1} and a pH (0.01 M $CaCl_2$) of 4.2; it then declined to reach a plateau. The SEBC was quartically related to X_m and cubically related to readily available Cu. The Cu sorption and desorption in organic soils can thus be assessed from easily determined properties such as SEBC and pH.

REFERENCES

Abdul-Halim, A.L. et al. 1981. An EPR spectroscopic examination of heavy metals in humic and fulvic acid soil fractions. *Geochim. Cosmochim. Acta.,* 45:481–487.

Alberts, J.J. and Giesy, J.P. 1983. Conditional stability constants of trace metals and naturally occurring humic materials: application in equilibrium models and verification with field data, in *Aquatic and Terrestrial Humic Materials.* Christman, R.F. and Gjessing, E.T., Eds., Ann Arbor Science, Ann Arbor, Michigan, 333–348.

Basta, N.T. and Tabatabai, M.A. 1992a. Effect of cropping systems on adsorption of metals by soils, I. Single-metal adsorption. *Soil Sci.,* 153:108–114.

Basta, N.T. and Tabatabai, M.A. 1992b. Effect of cropping systems on adsorption of metals by soils, III. Competitive adsorption. *Soil Sci.,* 153:331–337.

Basu, A.N., Mukherjee, D.C., and Mukherjee, S.K. 1964. Interaction between humic acid fraction of soil and trace element cations. *J. Indian Soc. Soil Sci.,* 12:311–318.

Bear, F.E. 1957. Toxic elements in soils, in *Soils: The Yearbook of Agriculture.* U.S. Government Printing Office, Washington, 165–172.

Beckwith, R.S. 1959. Titration curves of soil organic matter. *Nature,* 184:745–746.

Bohn, H.L., McNeal, B.L., and O'Connor, G.A. 2001. *Soil Chemistry,* 3rd ed. Wiley-Interscience, New York.

Bowen, H.J.M. 1966. *Trace Elements in Biochemistry.* Academic Press, New York.

Boyle, R.W. 1977. Cupriferrous bogs in the Sackville area, New Brunswick, Canada. *J. Geochem. Explor.,* 8:495–527.

Brennan, R.F., Robson, A.D., and Gartrell, J.W. 1983. Reactions of copper with soil affecting its availability to plants, II. Effect of soil pH, soil sterilization and organic matter on the availability of applied copper. *Aust. J. Soil Res.,* 21:155–163.

Brennan, R.F., Gartrell, J.W., and Robson, A.D. 1980. Reactions of copper with soil affecting its availability to plants, I. Effect of soil type and time. *Aust. J. Soil Res.,* 18:447–459.

Bunzl, K., Schmidt, W., and Sansoni, B. 1976. Kinetics of ion exchange in soil organic matter, IV. Adsorption and desorption of Pb^{2+}, Cu^{2+}, Cd^{2+}, Zn^{2+} and Ca^{2+} by peat. *J. Soil Sci.,* 27:32–41.

Cavallaro, N. and McBride, M.B. 1978. Copper and cadmium adsorption characteristics of selected acid and calcareous soils. *Soil Sci. Soc. Am. J.,* 42:550–556.

Christensen, J.B. et al. 1998. Proton binding by groundwater fulvic acids of different age, origins, and structure modeled with the model V and NICA-Donnan model. *Environ. Sci. Technol.,* 32:3346–3355.

CPVQ. 1996. *Crop Fertilization Guide* (in French). Agdex 540, 2nd ed., Conseil des Productions Végétales du Québec, Québec, Canada.

Dhillon, S.K., Sidhu, P.S., and Sinha, M.K. 1981. Copper adsorption by alkaline soils. *J. Soil Sci.,* 32:571–578.

Donahue, R.L., Mille, R.W., and Shickluna, J.C. 1983. An introduction to soils and plant growth. 5th ed., Prentice-Hall, Englewood Cliffs, New Jersey.

Duquette, M. and Hendershot, W.H. 1990. Copper and zinc sorption on some horizons of Quebec soils. *Commun. Soil Sci. Plant Anal.,* 21:377–394.

Ellis, B.G. and Knezek, B.D. 1972. Adsorption reactions of micronutrients in soils, in *Micronutrients in Agriculture.* Mortvedt, J. J., Giordano, M.P., and Lindsay, W.L. (Eds.), Soil Sci. Soc. Am. Inc., Madison, WI, 59–78.

Goodman, B.A. and Cheshire, M.V. 1976. The occurrence of copper-porphyrin complexes in soil humic acids. *J. Soil Sci.,* 27:337–347.

Goodman, B.A. and Cheshire, M.V. 1973. Electron paramagnetic resonance evidence that copper is complexed in humic acid by the nitrogen of porphyrin groups. *Nature,* 244:158–159.

Hamilton, H.A. 1979. Copper availability and management in organic soils. *Canada Agriculture,* 24(3):28–29.

Harter, R.D. 1992. Competitive sorption of cobalt, copper, and nickel ions by a calcium saturated soil. *Soil Sci. Soc. Am. J.,* 56:444–449.

Harter, R.D. 1979. Adsorption of copper and lead by Ap and B$_2$ horizons of several northeastern U.S. soils. *Soil Sci. Soc. Am. J.,* 43:679–683.

Hickey, M.G. and Kittrick, A. 1984. Chemical partitioning of cadmium, copper, nickel and zinc in soils and sediments containing high levels of heavy metals. *J. Environ. Qual.,* 13:372–376.

Jarvis, S.C. 1981. Copper sorption by soils at low concentrations and relation to uptake by plants. *J. Soil Sci.,* 32:257–269.

Jasmin, J.J. and Hamilton, H.A. 1980. Vegetable production on organic soils in Canada, in *The Diversity of Peat —Peatlands Seminar.* Pollett, F.C., Rayment, A.F., and Robertson, A., Eds., Newfoundland & Labrador Peat Association, Memorial University Newfoundland, St. John's, Newfoundland, 51–57.

Kabata-Pendias, A. 2001. *Trace Elements in Soils and Plants.* 3rd ed., CRC Press, Boca Raton, FL.

Kadlec, R.H. and Rathbun, M.A. 1983. Copper sorption on peat, in *Proc. Int. Symposium on Peat Utilization.* Fuchsman, C.H. and Spigarelli, S.A., Eds., Bemidji State University, Bemidji, MN, 351–364.

Karam, A. 1993. Chemical properties of organic soils, in *Soil Sampling and Methods of Analysis.* Carter, M.R., Ed., Lewis Publ., Boca Raton, FL, 459–471.

Khan, S.U. 1969. Interaction between the humic acid fraction of soils and certain metallic cations. *Soil Sci. Soc. Amer. Proc.,* 33:851–854.

Kishk, F.M. and Hassan, M.N. 1973. Sorption and desorption of copper by and from clay minerals. *Plant Soil,* 39:497–505.

Leeper, G.W. 1978. *Managing the Heavy Metals on the Land.* Marcel Dekker Inc., New York.

Lévesque, M.P. and Mathur, S.P. 1988. Soil tests for copper, iron, manganese, and zinc in histosols: 3. A comparison of eight extractants for measuring active and reserve forms of the elements. *Soil Sci.,* 145:215–221.

Lévesque, M.P., and Mathur, S.P. 1986. Soil tests for copper, iron, manganese, and zinc in histosols: 1. The influence of soil properties, iron, manganese, and zinc on the level and distribution of copper. *Soil Sci.,* 142:153–163.

Lévesque, M.P. and Mathur, S.P. 1984. The effects of using copper for mitigating histosol subsidence on: 3. The yield and nutrition of minicarrots, carrots, and onions grown in histosols, mineral sublayers, and their mixtures. *Soil Sci.,* 138:127–137.

Lévesque, M.P. and Mathur, S.P. 1983a. Effect of liming on yield and nutrient concentration of reed canarygrass grown in two peat soils. *Can. J. Soil Sci.,* 63:469–478.

Lévesque, M.P. and Mathur, S.P. 1983b. The effects of using copper for mitigating histosol subsidence on: 2. The distribution of copper, manganese, zinc, and iron in an organic soil mineral sublayers, and their mixtures in the context of setting a threshold of phytotoxic soil-copper. *Soil Sci.,* 135:166–176.

Lucas, R.E. 1982. Organic soils (histosols). Formation, distribution, physical and chemical properties and management for crop production. Res. Rep. No. 435, Farm Science, Michigan State University, East Lansing, MI.

MacLean, A.J. et al. 1964. Comparison of procedures for estimating exchange properties and availability of phosphorus and potassium in some eastern Canadian organic soils. *Can. J. Soil Sci.,* 44:66–75.

Mathur, S.P. 1983. A lack of bactericidal effect of subsidence-mitigating copper in organic soils. *Can. J. Soil Sci.,* 63:645–649.

Mathur, S.P. 1982a. The inhibitory role of copper in the enzymic degradation of organic soils, in *Proc. Int. Symposium on Peat Utilization.* Fuchsman, C.H. and Spigarelli, S.A., Eds., Bemidji State University, Bemidji, MN, 191–219.

Mathur, S.P. 1982b. Organic soil subsidence: A scan of conventional wisdom and current research, in *Proc. Organic Soil Mapping Workshop.* Land Resource Research Institute, Agriculture Canada, Fredericton, New Brunswick, Canada, 139–156.

Mathur, S.P., Hamilton, H.A., and Preston, C.M. 1979a. The influence of variation in copper content of an organic soil on the mineral nutrition of oats grown *in situ. Commun. Soil Sci. Plant Anal.,* 10:1399–1409.

Mathur, S.P., Hamilton, H.A., and Lévesque, M.P. 1979b. The mitigating effect of residual fertilizer copper on the decomposition of an organic soil *in situ. Soil Sci. Soc. Am. J.,* 43:200–203.

Mathur, S.P. and Lévesque, M.P. 1983. The effects of using copper from mitigating histosol subsidence on: 2. The distribution of copper, manganese, zinc, and iron in an organic soil, mineral sublayers, and their mixtures in the context of setting a threshold of phytotoxic soil-copper. *Soil Sci.,* 135:166–176.

Mathur, S.P. and Lévesque, M.P. 1989. Soil tests for copper, iron, manganese, and zinc in histosols: 4. Selection on the basis of soil chemical data and uptakes by oats, carrots, onions, and lettuce. *Soil Sci.,* 148:424–432.

Mathur, S.P., Lévesque, M.P., and Sanderson, R.B. 1989. The influence of soil properties, total copper, iron, manganese and copper on the yield of oat, carrot, onion and lettuce. *Commun. Soil Sci. Plant Anal.,* 20:1809–1820.

Mathur, S.P., MacDougall, J.I., and McGrath, M. 1980. Levels of activities of some carbohydrates, protease, lipase, and phosphatase in organic soils of differing copper content. *Soil Sci.,* 129:376–385.

Mathur, S.P. and Rayment, A.F. 1977. Influence of trace element fertilization on the decomposition rate and phosphorus activity of a mesic fibrisol. *Can. J. Soil Sci.,* 57:397–408.

Mathur, S.P. and Sanderson, R.B. 1978. Relationships between copper contents, rates of soil respiration and phosphatase activities of some histosols in an area of southwestern Quebec in the summer and the fall. *Can. J. Soil Sci.,* 58:125–134.

Matsuda, K. and Ikuta, M. 1969. Adsorption strength of zinc for soil humus, I. Relationship between adsorption forms and adsorption strengths of zinc added to soils and soil humus. *Soil Sci. Plant Nutr.,* 15:169–174.

McBride, M.B. 2000. Chemisorption and precipitation reactions, in *Handbook of Soil Science.* Sumner, M.E., Ed., CRC Press, Boca Raton, FL, B265-B302.

McBride, M.B. 1994. *Environmental Chemistry of Soils.* Oxford University Press, New York.

McBride, M.B. and Blasiak, J.J. 1979. Zinc and copper solubility as a function of pH in an acid soil. *Soil Sci. Soc. Am. J.,* 43:866–870.

McKeague, J.A. 1978. *Manual on Soil Sampling and Methods of Analysis,* 2nd ed., Canadian Society of Soil Science, Ottawa, Ontario.

McKenzie, R.M. 1980. The adsorption of lead and other heavy metals on oxides of manganese and iron. *Aust. J. Soil Res.,* 18:61–73.

McLaren, R.G., Swift, R.S., and Williams, J.G. 1981. The adsorption of copper by soil materials at low equilibrium solution concentrations. *J. Soil Sci.,* 32:247–256.

McLaren, R.G. and Crawford, D.V. 1973. Studies on soil copper, II. The specific adsorption of copper by soils. *J. Soil Sci.,* 24 443–452.

Mengel, D. and Rehm, G. 2000. Fundamentals of fertilizer application, in *Handbook of Soil Science.* Sumner, M.E., Ed., CRC Press, Boca Raton, FL, D155-D174.

Mortvedt, J.J. 2000. Bioavailability of micronutrients, in *Handbook of Soil Science.* Sumner, M.E., Ed., CRC Press, Boca Raton, FL, D71-D112.

Murray, K. and Linder, P.W. 1983. Fulvic acids: structure and metal binding, I. A random molecular model. *J. Soil Sci.,* 34:511–523.

Page, A.L., Miller, R.H., and Keeney, D.R. 1982. *Methods of Soil Analysis, Part 2. Chemical and Microbiological Properties,* 2nd ed., Agronomy no 9. Soil Sci. Soc. Am. Inc., Madison, WI.

Parent, L.E. and Perron, Y. 1983. Copper sorption by three peat types (in French with English summary). *Naturaliste Can.,* 110:67–70.

Petruzzelli, G., Guidi, G., and Lubrano, L. 1981. Influence of organic matter on lead adsorption by soil. *Zeitschrift für Pflanzenernährung und Bodenkunde,* 144:74–76.

Preston, C.M., Mathur, S.P., and Rauthan, B.S. 1981. The distribution of copper, amino compounds, and humus fractions in organic soils of different copper content. *Soil Sci.,* 131:344–352.

Rachid, M.A. 1974. Absorption of metals on sedimentary and peat humic acids. *Chem. Geol.,* 13:115–123.

Ravat, C., Monteil-Rivera, F., and Dumonceau, J. 2000. Metal ions binding to natural; organic matter extracted from wheat bran: Application of the surface complexation model. *J. Colloid Interface Sci.,* 225:329–339.

Sapek, B. 1976. Study on the copper sorption kinetics by peat-muck soils. *Proc. 5th Int. Peat Congr.,* II:236–245.

Sapek, B. and Zebrowski, W. 1976. Comparison of copper binding rate by peat-muck soils at various transformation stages. *Polish J. Soil Sci.,* IX:93–100.

SAS Institute, Inc. 1990. *SAS/STAT User's Guide. Version 6,* 4th ed. SAS Institute, Inc. Cary, NC.

Schnitzer, M. 1969. Reactions between fulvic acid, a soil humic compound and inorganic soil constituents. *Soil Sci. Soc. Amer. Proc.,* 33:75–81.

Schnitzer, M. 1978. Humic substances: Chemistry and reactions, in *Soil Organic Matter.* Schnitzer, M. and Khan, S.U., Eds., Elsevier, New York, 1–64.

Schnitzer, M. and Khan, S.U. 1972. *Humic Substances in the Environment.* Marcel Dekker Inc., New York.

Schnitzer, M. and Skinner, S.I.M. 1963. Organo-metallic interactions in soil: 1. Reactions between a number of metal ions and the organic matter of a podzol B_h horizon. *Soil Sci.,* 96:86–93.

Senesi, N. et al. 1989. Chemical properties of metal-humic acid fractions of a sewage sludge-amended aridisol. *J. Environ. Qual.,* 18:186–194.

Shotyk, W., Nesbitt, H.W. and Fyfe, W.S. 1992. Natural and anthropogenic enrichments of trace metals in peat profiles. *Int. J. Coal Geol.,* 20:49–84.

Sposito, G. 1984. *The Surface Chemistry of Soils.* Oxford University Press, New York.

Sposito, G., Lund, L.J., and Chang, A.C, 1982. Trace metal chemistry in arid-zone field soils amended with sewage sludge: I. Fractionation of Ni, Cu, Zn, Cd, and Pb in solid phases. *Soil Sci. Soc. Am. J.,* 46:260–264.

Stevenson, F.J. 1982. *Humus Chemistry. Genesis, Composition, Reactions.* Wiley-Interscience, New York.

Szalay, A., Sámsoni, Z., and Szilágyi, M. 1975. Manganese and copper deficiency of plants as a characteristic defect of lowmoor peat soils. *Zeitschrift für Pflanzenernährung und Bodenkunde*, 138:447–458.

Szalay, A. and Szilágyi, M. 1968. Laboratory experiments on the retention of micronutrients by peat humic acids. *Plant Soil,* 29:219–224.

Tan, K.H. 1998. *Principles of Soil Chemistry,* 3rd ed., Marcel Dekker Inc., New York.

Taylor, S.R. 1965. The application of trace element data to problems in petrology, in *Physics and Chemistry of the Earth.* Aherns, L.H. et al., Eds., Pergamon Press, New York, 133–213

Tessier, A., Campbell, P.G.C., and Bisson, M. 1979. Sequential extraction procedure for the speciation of particulate trace metals. *Anal. Chem.,* 51:844–850.

Tyler, L.D. and McBride, M.B. 1982. Mobility and extractability of cadmium, copper, nickel, and zinc in organic and mineral soil columns. *Soil Sci.,* 134:198–205.

Veith, J. A. and Sposito, G. 1977. On the use of the Langmuir equation in the interpretation of "adsorption" phenomena. *Soil Sci. Soc. Am. J.,* 41 697–702.

Verloo, M., Kiekens, L., and Cottenie A. 1973. Experimental study of Zn and Cu mobility in the soil. *Mededelingen Faculteit Landbouwwetenscchappen,* 38:380–388.

Viets, F.G. 1962. Chemistry and availability of micronutrients in soils. *J. Agric. Food Chem.,* 10:174–177.

CHAPTER **8**

Fate of Pesticides in Organic Soils

Josée Fortin

CONTENTS

ABSTRACT

    This chapter reviews the current knowledge on pesticides fate when applied to organic soils. Pesticide retention, the process controlling the fate and persistence of pesticides in soils, depends on key pesticides properties such as polarity and hydrophobicity, and on soil organic matter quantity and quality. Generally, pesticide retention is higher in soils where the organic matter is in a more decomposed stage, although contradictory results are reported. Pesticide retention by organic soils can be irreversible and produce bound residues. The long-term fate and environmental importance of bound residues in organic soils is unknown. Some processes that decrease pesticide concentration in organic soils, such as plant uptake, degradation, erosion, and leaching, are discussed. The overall persistence of pesticides in surface horizons is higher in organic than in mineral soils. This persistence is usually related to pesticide retention by soil components and can result in soil accumulation of some pesticides with time. The pesticides applied to organic soils can affect biochemical processes and microbial activities. Generally, the results reported show that the effects do not persist for a long period of time.

I. INTRODUCTION

Most pesticides interact with soil organic matter (Turco and Kladivko, 1994; Stevenson, 1985, 1994; Weber, 1994), and this retention controls their bio-availability, leaching, degradation and volatilization in organic soils used for vegetable crop production. To obtain the same level of pest control, soil-applied pesticides are usually recommended at higher rates in organic than in mineral soils (CPVQ, 1997; Khan et al., 1976a; Stevenson, 1985). The reasons are:

1. Poor pesticide bio-activity due to retention by soil humus (Jourdan et al., 1998)
2. Higher water content in organic soils on a volume basis compared with mineral soils, so more solute is required in the former to achieve equal and effective pesticide concentrations (Mathur and Farnham, 1985)

To increase linuron efficiency in organic soils, it is sometimes recommended to water the soil prior to pesticide application, thus reducing pesticide retention and increasing bio-availability (CPVQ, 1997). The use of certain pesticides is forbidden in organic soils but not in mineral soils. Trifluralin is so strongly retained by soil organic matter and is so persistent in organic soils, even in its inactivated form (Braunschweiler, 1992), that it can only be used on mineral soils with low organic matter content (CPVQ, 1997).

Higher application rates of pesticides on organic soils leads to their accumulation and persistence (Khan et al., 1976a). Some pesticides can be released slowly from humus by microbes (Hsu and Bartha, 1974; Mathur and Morley, 1975; Khan, 1982), taken up by mature crops (Morris and Penny, 1971; Khan et al., 1976a, 1976b; Bélanger and Hamilton, 1979), and contributive to pest resistance (Suett, 1975) or disturbance of desirable microbial activities (Mathur et al., 1980a).

Although the importance of soil organic matter on pesticides behavior in soils is well recognized, no synopsis of the behavior and fate of pesticides applied to cultivated organic soils is available. The aim of this chapter is to review the different aspects related to the behavior of pesticides used for pest control in crops grown in organic soils, and to summarize the present knowledge on pesticide–organic soils interactions.

II. PESTICIDE RETENTION BY SOIL ORGANIC MATTER

Pesticide retention by soil is one of the main factors affecting the pollution potential of a pesticide because it controls its concentration in soil solution and its biological availability, persistency, and mobility (Franco et al., 1997). Retention is affected by pesticides properties as well as soil organic matter (SOM) content and composition.

A. Pesticide Properties Affecting Their Retention by Soils

Most pesticides used today are organic synthetic molecules. Table 8.1 presents selected properties of some pesticides. The chemical structure of the organic molecule determines the properties that control its behavior in the environment. When considering their interactions with soil constituents, pesticides can be grouped into polar and nonpolar molecules. A polar molecule has positive and negative poles,

Table 8.1 Selected Properties of Some Pesticides

Common Name	Chemical Family	pK_A	Log (K_{ow}) $\left[\dfrac{(mol \cdot L \text{ octanol}^{-1})}{mol \cdot L \text{ water}^{-1}}\right]$	Water Solubility $(mg\ L^{-1})$
Nonionic				
1,3-D	Organochlorine	—	2.28	2.7×10^3
Chlorpyrifos	Organophosphate	—	4.7–5.3	0.45–1.30
Cypermethrin	Pyrethroid	—	6.60	0.004
Linuron	Substituted urea	—	3.00	55–81
Metolachlor	Acetamide	—	2.60–3.28	488–550
Monolinuron	Substituted urea	—	2.20	580–735
Permethrin	Pyrethroid	—	6.10	0.006–0.2
Thiobencarb	Thiocarbamate	—	3.42	30
Trifluralin	Dinitroaniline	—	3.97–5.33	< 1
Acidic				
Sethoxydim	Cyclohexene oxime	4.6	4.51 (pH 5) 1.65 (pH 7)	25 (pH 4) 4700 (pH 7)
Basic				
Metamitron	Triazinone	n.a.	0.8	20,000
Metham sodium	Thiocarbamate	17.6	< 1.00	$> 7.2 \times 10^4$
Paraquat	Bipyridillium	11	−4.5	7.0×10^5
Prometryn	Triazine	4.05	3.1	33–46
Amphoteric				
Glufosinate ammonium	Organophosphate	$pK_{A1} = < 2$ $pK_{A2} = 2.9$ $pK_{A3} = 9.8$	< 0.1	1.37×10^6
Glyphosate	Organophosphate	$pK_{A1} = < 2$ $pK_{A2} = 2.6$ $pK_{A3} = 5.6$ $pK_{A4} = 10.6$	−1.6	1.2×10^4

Sources: From ARS Pesticide Properties, www.arsusda.gov/acsl/ppdb.html and Weber, J. B. 1994. Properties and behavior of pesticides in soil, in *Mechanisms of Pesticide Movement into Groundwater*. Honeycutt, R.C. and Schabacker, D.J., Eds., Lewis Publishers, Boca Raton, FL, 15–42. With permission.

which are areas where a partial charge occurs along the molecule. The polarity of a molecule is due to a combination of polar bonds (bonds between ions of different electronegativity) and molecular geometry. As a rule of thumb, one can check to see if certain highly electronegative ions, such as F, O, N, Cl, and Br, are present on the pesticide chemical structure and if they form nonsymmetrical bonds. In such a case, the molecule is probably polar. The polar character of an organic molecule influences its water solubility and its affinity with soil components, such as clay minerals and organic matter. A polar pesticide will usually have a higher water solubility than a nonpolar one. It is also held by both clay minerals and SOM, while nonpolar pesticides are retained almost exclusively by SOM (Stevenson, 1994).

Polar molecules can be nonionic, ionic, or ionizable. Ionization potential of a molecule depends on the functional groups of its chemical structure and on the relative position of those groups. Basic molecules can accept protons and become positively charged due to basic functional groups such as –NH- and NH$_2$. Acidic molecules can give up protons, with the development of negative charges on their structures. The main acidic functional groups found on pesticides are –COOH and –OH. The proportion of charged and neutral molecules for either basic or acidic pesticides depends on their acid dissociation constant (K$_A$) and the ambient pH. For an acidic pesticide, the relationship is:

$$\log \frac{[A^-]}{[HA]} = pH - pK_A \qquad (8.1)$$

where HA is the acid pesticide, A$^-$ is the conjugate base of HA, and pK$_A$ is defined as –log(K$_A$). To compare acids and bases on a uniform scale, we can obtain a similar relationship for basic pesticides using the acidity constant of the conjugate acid (BH$^+$). The relationship obtained is:

$$\log \frac{[B]}{[BH^+]} = pH - pK_A \qquad (8.2)$$

where B is the basic pesticide and BH$^+$ is the conjugate acid. Some pesticides have both acidic and basic functional groups and are said to be amphoteric. Glyphosate is the typical example of an amphoteric pesticide, with three acidic functional groups and one basic functional group. Finally, nonionic molecules do not develop positive or negative charges. The charge characteristic of a pesticide influences not only its retention by soil components (Stevenson, 1994; Weber, 1994), but also its absorption by plants (Sterling, 1994), water solubility (Weber, 1994), and, indirectly, degradation kinetics.

Polarity and ionization potential influence the hydrophobic character of a molecule, which is usually quantified for an organic molecule by the octanol–water partition coefficient (K$_{ow}$). The K$_{ow}$ for a given compound is defined as the ratio of its concentration in octanol (an amphiphilic organic solvent historically chosen to represent the natural organic matter) over its concentration in water at equilibrium. The higher the K$_{ow}$ value of a pesticide, the more hydrophobic it is. Because SOM is the only soil constituent with hydrophobic character, hydrophobic pesticides are mainly attracted by SOM. This attraction will depend on SOM composition, which in turn depends on the original material and its degree of decomposition. A general direct inverse relationship exists between the K_{ow} of a given pesticide molecule and its water solubility (Schwarzenbach et al., 1993). For nonpolar pesticides, the K_{ow} is usually directly proportional to the adsorption potential of the molecule by soils, which is described using partition coefficients such as K_d (soil/water partition coefficient), K_{om} (organic matter/water partition coefficient) and K_{oc} (organic carbon/water partition coefficient). These coefficients are discussed in detail next.

B. Soil Organic Matter Composition

The nature of pesticide–soil interactions depends not only on the type of pesticide, but also on soil composition (Almendros, 1995). Organic soils composed of organic material at different decomposition stages are expected to react differently with pesticides, mainly because the polarity of the organic matter will vary (Torrents et al., 1997). Due to the complex nature of SOM, its polarity has been described using a polar to nonpolar group ratio, which is the ratio of the sum of its nitrogen and oxygen contents over its carbon content [$(N + O)/C$] (Rutherford et al., 1992). This ratio is higher for relatively fresh compared with more decomposed materials (Torrents et al., 1997). The rubbed fiber content (RF) and the pyrophosphate index (PI) are used to describe the degree of peat decomposition (Morita, 1976). These indices classify organic soils from fibric to sapric materials (Soil Classification Working Group, 1998; Soil Survey Staff, 1990). The pyrophosphate index is evaluated on chromatographic paper using a Munsell color chart (Soil Classification Working Group, 1998; Soil Survey Staff, 1990), or from absorbance of a sodium pyrophosphate extract of the soil at 550 nm (Kaila, 1956; Vaillancourt et al., 1999). The higher the pyrophosphate index, the more humified is SOM.

C. Mechanisms of Pesticide Retention by SOM

The main binding mechanisms involved in the interaction of pesticides with SOM (e.g., Van der Waals forces, hydrophobic attractions, ionic exchanges, ligand exchanges, and charge-transfer complexes) are described in detail in Chiou (1990), Senesi (1992), and Stevenson (1994). The acting mechanism depends on the pesticide involved and on sorbent composition. Ionic exchange is possible only for ionized basic and acidic pesticides, while the charge-transfer mechanism can take place for electron donor pesticides such as s-triazines and substituted ureas (Senesi and Chen, 1989; Senesi and Miano, 1995). Nonpolar pesticides are retained mainly by hydrophobic attraction on hydrophobic constituents of SOM (Torrents et al., 1997) following a process called partitioning. Several mechanisms may also be involved. For example, bipyridillium pesticides, such as diquat and paraquat, are retained through cation exchange and can form charge-transfer complexes with aromatic constituents of humic substances (Khan, 1973). Sorption mechanisms can provide information on strength and reversibility of pesticide-soil interactions, and on soil constituents (e.g., SOM, clay minerals, etc.) that can be involved in pesticide retention.

D. Quantitative Description of Pesticide Retention by SOM

We are often interested in the amount of a given compound retained under certain conditions. The conventional method used to evaluate pesticide retention is to construct sorption isotherms, where the amount of a pesticide retained by a soil is related to pesticide equilibrium concentration in solution. Different models can be used to describe the isotherms obtained. For pesticide sorption, the Freundlich equation is used extensively because it allows the description of different forms of isotherms. The Freundlich equation can be written as follows:

$$C_a = K_f C_w{}^n \qquad (8.3)$$

where C_a is the amount of sorbate (mg kg^{-1}), C_w is pesticide concentration in solution at equilibrium (mg L^{-1}), and K_f (L kg^1) and n (unitless) are constants obtained by curve fitting of experimental data. Retention of nonpolar pesticides by SOM is often linear (n = 1), and the Freundlich coefficient (K_f) is then called the partition coefficient (K_d). The effect of SOM on pesticides retention can be separated from that of other factors using normalized soil adsorption coefficients. The first coefficient is the K_{om}, which is the ratio of K_f or K_d divided by SOM content. Using organic C, we obtain the K_{oc} (K_{oc} = 100 [(K_f or K_d)/(organic C)]). These normalized soil adsorption coefficients are often interpreted as a measure of the contribution of hydrophobic forces to adsorption. Whereas this would be true for nonpolar compounds, this interpretation cannot be used for polar species for which retention mechanisms other than hydrophobic forces may predominate (Franco et al., 1997). Furthermore, the qualitative differences in SOM composition may also affect polar and nonpolar pesticides retention (Rutherford et al., 1992; Kile et al., 1995). These factors contribute to the variability of sorption coefficients reported in the literature for polar as well as nonpolar pesticides.

Table 8.2 lists some examples of sorption coefficients evaluated for some polar pesticides on different organic sorbents. Most pesticides show increased sorption with increasing degree of decomposition of the organic sorbent (Morita, 1976; Braverman et al., 1990a; Franco et al., 1997; Torrents et al., 1997) The only contrasting results are the ones reported by Parent and Bélanger (1985), who found that linuron retention was higher on hemic than on sapric soil material. They explained their contrasting results by the high ash content of the peat material used in their study (from 6 to 7% for the hemic material and from 14 to 37% for the sapric material), because linuron sorption was negatively related to ash content. Linuron is a nonionic polar pesticide that can be retained differentially by organic and mineral materials. The results of Braverman et al. (1990a) (Table 8.2) show that thiobencarb sorption by two moorsh soils was greater than on sand on a whole soil basis (K_d, K_f, or %); however, the sand adsorbed a greater amount of the pesticide per C unit (K_{oc}). The greater activity of organic C in the sand was attributed to its more advanced state of decay of the parent material. As SOM decomposes, the humic fraction, which is the most reactive SOM fraction, increases. Furthermore, in soils with high organic C contents, SOM may be aggregated into more compact grains, resulting in a decrease in available adsorptive surface per unit weight of organic C. The contri-

Table 8.2 Sorption Coefficient of Some Herbicides Evaluated on Organic and Mineral Soils

Pesticide Common Name	K_f (cm^3 g^{-1})	n	K_{oc} (cm^3 g^{-1})	Soil Type/Conditions	Reference
Linuron				**Fibric Soil Material**	
	24	0.92	—	Fibric (RF = 84%, PI = 8)	Morita (1976)
	158	0.77	—	Fibric (RF = 76%, PI = 5)	Morita (1976)
	269.15	0.62	651.4	Sphagnofibrist, 41.3% C_{org}	Franco et al. (1997)
				Hemic Soil Material	
	196	0.79	—	Hemic (RF = 14%, PI = 4)	Morita (1976)
	174	0.69	—	Hemic (RF = 30%, PI = 7)	Morita (1976)
	295.12	0.74	541.3	Hemic, 6% ash	Parent and Bélanger (1985)
	257.04	0.94	476.5	Hemic, 7% ash	Parent and Bélanger (1985)
				Sapric Soil Material	
	297	0.67	—	Sapric (RF = 10%, PI = 0)	Morita (1976)
	231	0.73	—	Sapric (RF = 10%, PI = 1)	Morita (1976)
	89.13	1.37	178.7	Sapric, 14% ash	Parent and Bélanger (1985)
	54.95	1.18	150.4	Sapric, 37% ash	Parent and Bélanger (1985)
	371.53	0.53	852.1	Sapric, 43.6% C_{org}	Franco et al. (1997)
Metamitron	23.99	0.50	58	Sphagnofibrist, 41.3% C_{org}	Franco et al. (1997)
	54.95	0.38	126	Sapric, 43.6% C_{org}	Franco et al. (1997)

Table 8.2   Sorption Coefficient of Some Herbicides Evaluated on Organic and Mineral Soils   *(Continued)*

Pesticide Common Name	K_f (cm³ g⁻¹)	n	K_{oc} (cm³ g⁻¹)	Soil Type/Conditions	Reference
Thiobencarb	339	0.94	765	Moorsh (Medisaprist), 48.6% C_{org}	Braverman et al. (1990a)
	169	0.92	539	Moorsh (Medihemist), 34.1% C_{org}	Braverman et al. (1990a)
	14	0.95	1195	Fine sans (Haploquod), 1.1% C_{org}	Braverman et al. (1990a)
Metolachlor	460	1.00	800	Lignin, (O + N)/C = 0.537	Torrents et al. (1997)
	225	1.00	402	Collagen, (O + N)/C = 0.634	Torrents et al. (1997)
	3.2	1.00	7.2	Chitin, (O + N)/C = 1.01	Torrents et al. (1997)
	3.5	1.00	7.9	Cellulose, (O + N)/C = 1.11	Torrents et al. (1997)
Alachlor	402	1.00	704	Lignin, (O + N)/C = 0.537	Torrents et al. (1997)
	2574	1.00	459	Collagen, (O + N)/C = 0.634	Torrents et al. (1997)
	5.4	1.00	12.1	Chitin, (O + N)/C = 1.01	Torrents et al. (1997)
	6.45	1.00	14.5	Cellulose, (O + N)/C = 1.11	Torrents et al. (1997)
Propachlor	140	1.00	245	Lignin, (O + N)/C = 0.537	Torrents et al. (1997)
	22.9	1.00	40.9	Collagen, (O + N)/C = 0.634	Torrents et al. (1997)
	1.45	1.00	3.25	Chitin, (O + N)/C = 1.01	Torrents et al. (1997)
	0.57	1.00	1.28	Cellulose, (O + N)/C = 1.11	Torrents et al. (1997)

Note: RF = rubbed fiber content; PI = pyrophosphate index; C_{org} = organic carbon content.

bution of the mineral fraction of the sandy soil to adsorption is also a possible explanation for the difference, while adsorption by old-cultivated moorsh soils may be completely dependent on their organic C content (Braverman et al., 1990a).

E. Pesticide-Bound Residues

The initial reactions between pesticide and SOM are reversible. As reaction time between pesticide and SOM increases, however, the reversibility of sorption reactions often decreases with a concomitant decrease in pesticide extractability (Mathur and Morley, 1978). The pesticide fraction that remains strongly bound to soil particles following extraction is called bound residue. A review on bound residues in mineral soils can be found in Khan (1988). Pesticides-bound residues can only be quantified in the laboratory using ^{14}C-labeled parent compounds. Following reaction of the soil with the ^{14}C-labeled pesticide and pesticide extraction with an organic solvent, the amount of ^{14}C remaining in the soil residue is the pesticide bound residue, which can be hydrolyzed for quantification and identification. Various hypotheses have been proposed to explain bound residue formation, such as chemical binding to soil organic constituents, incorporation into phenolic polymers, bioincorporation into cellular structures through metabolic activity of soil microorganisms, and blocking of internal voids of SOM trapping the residue (Mathur and Morley, 1975; Bollag, 1992; Kästner et al., 1999). Most of the hypotheses include the fundamental role of soil organic constituents in the formation of bound residues.

Some evidences for the formation of bound residues of pesticides applied to organic soils can be found in the literature. In a study where ^{14}C-prometryn was applied to a hemist soil (sapric surface materials) and incubated in the laboratory for 150 days, Khan and Hamilton (1980) found that 43% of the initial ^{14}C added to the soil was in the form of bound residues. Using the same soil, Khan (1982) found that bound ^{14}C-labeled residues (57.4% of the radioactivity applied) following soil incubation with ^{14}C-prometryn for 1 year were associated with humin (57%), humic acid (10%) and fulvic acid (26%) fractions. Zhang et al. (1984) found that 19% of the ^{14}C applied with deltamethrin to an organic soil (decomposition stage unspecified) was in the form of bound residues after an incubation period of 180 days. Most of the ^{14}C was found to be bound to humin (58.5–65.6%), while humic and fulvic acids contained between 21.7–24.8% and 7.1–16.8% of the radioactivity, respectively. The bound residues associated with the low molecular weight or more soluble organic matter fraction (fulvic acids) may be considered as potentially bioavailable to both plants and exposed aqueous and soil fauna (Khan, 1982) and can potentially be mobile in soil. Braverman et al. (1990a) found that about 50% of the ^{14}C applied with thiobencarb to sapric and hemic soil materials remained as bound residue after 42 days. The greater amount of bound ^{14}C was present at the beginning of the experiment (7 and 14 days), indicating a rapid irreversible binding of thiobencarb by the soil in its original form. Bound residues can also be formed in organic soils from degradation products, as was shown for substituted urea (Hsu and Bartha, 1974), pyrethroid (Zhang et al., 1984), triazine (Khan, 1982) and chlorophenoxy (Scott et al., 1983; Hatcher et al., 1993) pesticides.

Although soil bound pesticide residues are very stable and may be in a form not harmful to the environment (Stevenson, 1994), they can be released with time. They can then be degraded (Khan and Ivarson, 1981), although their degradation rates can be much slower than those of initially applied pesticides (Raman and Rao, 1988). They can also be absorbed by plants, as shown by Khan (1980) for oat plants treated with ^{14}C-ring-labeled prometryn. The fate and toxicity of pesticide bound residues in organic soils remains uncertain, but the role of SOM in their formation is obvious.

III. PESTICIDE FATE IN ORGANIC SOILS

The persistence of a pesticide in soil depends greatly on processes acting to decrease its concentration at a given location in the soil environment. Volatilization from the soil surface is not an important process in organic soils because of the high retention of most pesticides by SOM (Chapman and Chapman, 1986). Other processes involved in pesticides dissipation are crop uptake, degradation, erosion, and leaching.

A. Crop Uptake

The crops treated with pesticide can absorb the molecules to various extents, depending on pesticide formulation, application method (foliar, soil), pesticide persistence, etc. Once absorbed by the plant, the pesticide can be degraded, translocated to, or accumulated in different tissues. The crop–pesticide interaction is crop specific and is one important factor considered before a pesticide can be homologated. In fact, the time required between the last pesticide application and the crop harvest depends greatly on the amount of pesticide that is tolerated in the mature crop, which is called the maximum residue limit (MRL). In Canada, MRL values are established for several compounds homologated for pest control in a given crop. In the case where no MRL has been established, the limit is set at 0.1 mg kg^{-1}, which is less than most values established, and can be considered as a general safety limit.

Some field studies examined pesticide residues in crops grown on pesticide-treated organic soils. In a study on the behavior of two herbicides, linuron and paraquat, in a hemist soil with sapric surface materials, Khan et al. (1976a) found that no linuron residue was present in carrots at harvest when the herbicide was applied in the spring. The following year in the same soil, with no herbicide application, onions and lettuce contained detectable amounts of linuron, while no residue was found in carrots. Lettuce and onion grown on paraquat-treated soils showed negligible amounts of the herbicide the same year of application. This herbicide was highly persistent, however, so that long-term safety for crops following repeated use of paraquat in organic soils is uncertain (Khan et al., 1976a). Carrots and radishes grown on a moorsh soil and a Painfield sand treated with different insecticides showed different residual pesticides concentrations (Chapman and Harris, 1980, 1981, 1982). Chlorpyrifos residues in mature crops grown in the moorsh soil were lower (< 0.01 mg kg^{-1}) compared with those found in crops grown in the sand (0.03 mg kg^{-1} for carrots and 0.09 mg kg^{-1} for radishes), illustrating the lower pesticide bioavailability in the organic soil (Chapman and Harris, 1980). The crops absorbed low levels of 3,5,6-trichloro-2-pyridinol, a degradation product of chlorpyrifos. Concentrations less than 0.01 mg kg^{-1} for both crops and concentrations of 0.01 mg kg^{-1} for carrots and 0.06 mg kg^{-1} for radishes were found in crops grown on moorsh and sand, respectively. No residues of permethrin and cypermethrin were found in radishes and carrots grown on those soils, within a detection limit of 0.01 mg kg^{-1} (Chapman and Harris, 1981). Residues of isofenphos and isazophos in radishes and carrots did not exceed 0.04 mg kg,$^{-1}$ except for carrots grown the first year on isofenphos-treated sand, where 0.25 mg kg^{-1} was found (Chapman and Harris, 1982). Chapman et al. (1984) found that residues of ethion, fonofos, chlorpyrifos, chlorfenvinphos, and carbofuran applied as furrow granular treatments to control onion maggots were less than 0.01 mg kg^{-1} in onions at harvest. Thus, the pesticides tested were absorbed by certain crops grown on organic soils, but concentrations in crops at harvest were plant specific and usually much lower than their MRL values.

B. Pesticide Degradation

Pesticide degradation reactions can be chemically or biologically mediated. Biodegradation remains the main type of degradation reaction in most soils, including organic soils (Chapman et al., 1981; Cheah et al., 1998). Comparisons of the persistence of different pesticides in natural and sterile organic soils have demonstrated the importance of biodegradation in pesticide dissipation (Chapman et al., 1985; Miles et al., 1981; Murty et al., 1982).

The SOM has the potential of promoting the nonbiological degradation of many organic pesticides (Stevenson, 1976). Pesticides adsorbed to both organic and inorganic soil constituents will have long half-lives in most soils, as is the case for paraquat (Cheah et al., 1998). Other pesticides which adsorb to SOM have longer half-lives in organic than in mineral soils.

In controlled laboratory experiments, Hitchings and Roberts (1980) observed that the time for depleting 50% of the applied flamprop-methyl was 1–2 weeks for a sandy loam, a clay loam, as well as a loam soil, and 2–3 weeks for an organic soil (characteristics unspecified). Chapman et al. (1981), who studied the degradation of some pyrethroid insecticides in a sandy loam and an organic soil (decomposition stage unspecified), showed that following soil incubations of pesticides for 8 weeks, the amount of the original compound remaining in the organic soil was always higher than in the mineral soil. Cheah et al. (1998) showed that mineralization of glyphosate, as evaluated by the release of $^{14}CO_2$ from the marked molecule, was longer in a moorsh soil (half-life of 309 days) than in a sandy loam soil (half-life of 19 days). This was attributed to the high sorption capacity of the moorsh soil, rendering the pesticide inaccessible to microbial metabolism. On the other hand, Braverman et al. (1990a) observed a shorter degradation half-life of thiobencarb in two moorsh soils (16 and 18 days) as compared with a sandy soil (24 days), even if the pesticide was strongly sorbed by SOM. The greater amount of SOM in the moorsh soils may have favored microbes degrading thiobencarb, thereby shortening its half-life. Furthermore, the degradation of thiobencarb to metabolites without evolution of $^{14}CO_2$ indicated that it may be co-metabolized (Braverman et al., 1990a).

Generally, soils with a history of a given pesticide application result in a shorter pesticide half-life than soils exposed to a given pesticide for the first time due to the adaptation of soil microflora (Leistra and Green, 1990). Cheah et al. (1998) attributed the longer mineralization half-life of 2,4-D in a sandy loam (36 days) as compared with a moorsh soil (3 days) to the fact that the former had no history of pesticide application while the latter received the pesticide for many years. The microflora adaptation to a pesticide molecule can be relatively rapid. Chapman et al. (1986) showed that the microflora of 3 mineral soils (a sandy loam, a sand and a clay loam) and one moorsh soil can develop anti-carbofuran activity within 28 days of an initial treatment; however, the carbofuran concentration to induce this activity was 10 times higher in the organic soil compared with the mineral soils.

Very few studies looked at the effects of soil factors or pesticide application methods on pesticide degradation in organic soils. Miles et al. (1984) found that the disappearance of two organophosphate insecticides, chlorpyrifos and chlorfenvinphos, was proportional to moisture content in a moorsh soil, while it was constant in a sandy soil. Sahid and Teoh (1994) compared the effects of soil moisture and temperature on the dissipation of terbuthylazine, a triazine herbicide, in a sandy loam and an organic soil (55% organic C, pH 3.1, unspecified decomposition stage) in a controlled, closed laboratory experiment. Assuming a first-order dissipation, the half-life of terbythylazine was shorter in the organic soil than in the sandy loam, especially at high temperatures. In the organic soil, dissipation half-life was shorter with increasing water content. Chapman and Chapman (1986) examined the effect of chlorpyrifos formulation on its dissipation in a moorsh soil and a Plainfield sand under controlled laboratory conditions. In the mineral soil, formulation only had a slight effect, while in the organic soil, chlorpyrifos applied as granular formulation disappeared at a slower rate than the one applied as an emulsifiable concentrate. This may reflect the effect of the physical form of the pesticide on its retention by SOM, which affected pesticide degradation.

C. Wind and Water Erosion

Organic soils are susceptible to wind and water erosion, although those phenomena are not so well documented. According to Wall et al. (1995), organic soils of southern Quebec are among the areas in Canada most vulnerable to wind erosion. Pesticides can be strongly retained in organic soils, therefore, they can be lost with eroded soil particles. Studies on mineral soils have revealed that application method (surface applied or incorporated) can affect the amount of pesticides loss by wind erosion. In a study conducted on a clay loam soil in Alberta, Canada, Larney et al. (1999) found that the overall wind erosion losses (expressed as percent of amount applied) of two soil-incorporated herbicides (average loss of 1.5%) were about three times lower than those of four surface applied herbicides (average loss of 4.5%). Loss of pesticides with runoff water and water eroded sediments is well documented for mineral soils (Leonard, 1990; Triegel and Guo, 1994) and can probably occur under certain conditions in organic soils. Soil erosion by wind and water thus represents potential pathways for environmental transport to off-target locations of pesticides applied to organic soils and should be investigated in the future.

D. Leaching and Colloid Facilitated Transport

High sorption potential of organic soils reduces considerably pesticides leaching through the soil matrix. Consequently, very few studies have been conducted on the subject. Fadayomi and Warren (1977) reported that oxyfluorfen was held tightly against desorption on moorsh soils, and less than 2% of the parent material was found in the leachate. Murty et al. (1982) also showed that nitrofen leaching in a moorsh soil was negligible. Braverman et al. (1990b) studied the mobility of thiobencarb, a very strongly sorbed pesticide, in two moorsh (sapric and hemic) materials and a sandy soil. They found that the mobility of this herbicide was slightly greater in sandy than in organic soils, but that most of the herbicide remained in the top 1 cm across soil types. Frank et al. (1991) found no residues of oxyfluorfen in tile drainage water from a treated area of an organic soil (characteristics unspecified) cultivated with onions on the Kettelby Research Station on the Holland marsh north of Toronto, Ontario.

Even if pesticides are strongly retained in organic soils, contamination of shallow groundwater by pesticides is possible through preferential flow and colloid-facilitated transport. Field evidences of preferential flow of solutes in organic soils are not available; however, Bergström (1995) conducted a column laboratory transport experiment on dichlorprop in undisturbed clay and organic (88.5% SOM, unspecified decomposition stage) soils. Early breakthrough of the pesticide in drainage water located 1.18 m below the soil surface indicated a bypass of the soil matrix and thus preferential flow. On average, 88% of the dichlorprop load in drainage water, under a water regime representing an average rain season, was found in the first 10 mm of leachate. This rapid pesticide movement was typical of preferential flow.

Nonpolar and polar organic chemicals can bind to water soluble SOM (Pennington et al., 1991). Fulvic acid-bound pesticide residues are the form most susceptible to mobilization in soils because fulvic acid is considered as the dominant soluble organic fraction (Zhang et al., 1984). This SOM fraction can become mobile under certain conditions and promote downward pesticides movement in soils, with a predominant movement through preferential flow. This process is often referred to as "colloid-facilitated transport" of pesticides. Its importance under field conditions is unknown, and more research is needed to evaluate the importance of this phenomenon in organic soils.

IV. FIELD PERSISTENCE OF SELECTED PESTICIDES

Table 8.3 summarizes studies on pesticide persistence in organic soils and, in some case, in mineral soils. Evidence suggests that some pesticides can persist for more than one growing season in organic soils. This was demonstrated for paraquat, linuron (Khan et al., 1976a), and isofenphos (Chapman and Harris, 1982). When soil persistence of a pesticide is relatively long, it can accumulate over time. In the fall of 1977, Miles and Harris (1979) did a survey of carbofuran content (and its main degradation products) in organic soils from 22 Ontario farms with a history of carbofuran granular application. Nineteen of the 22 soils contained detectable

Table 8.3   Field Persistence of Selected Pesticides in Soils

Pesticide	Application	Soil Type	Field Persistence — Time	Field Persistence — Amount Remaining (% of Applied)	References
Linuron (H)	WP, 2.24 kg ha⁻¹	Hemist	15 months	21	Khan et al. (1976a)
	WP, 4.48 kg ha⁻¹	Hemist	15 months	14	Khan et al. (1976a)
Nitrofen (H)	EC, 2.02 kg ha⁻¹	Moorsh	16 weeks	15	Murty et al. (1982)
	EC, 2.02 kg ha⁻¹	Plainfield sand	16 weeks	2	Murty et al. (1982)
Paraquat (H)	S, 2.24 kg ha⁻¹	Hemist	125 days	83	Mathur et al. (1976)
	S, 1.12 kg ha⁻¹	Hemist	15 months	54	Khan et al. (1976a)
	S, 2.24 kg ha⁻¹	Hemist	15 months	50	Khan et al. (1976a)
Prometryn (H)	WP, 4.48 kg ha⁻¹	Organic soil (45.4% OMC, pH 5.2, KPI = 20)	30 days	37	Mathur et al. (1980a)
Chlorpyrifos (I)	EC, 3.4 kg ha⁻¹	Moorsh	24 weeks	13	Chapman and Harris (1980)
	EC, 3.4 kg ha⁻¹	Plainfield sand	24 weeks	5	Chapman and Harris (1980)
Cypermethrin (I)	EC, 280 g ha,⁻¹ incorporated	Moorsh	6 months	6	Chapman and Harris (1981)
	EC, 140 g ha,⁻¹ not incorporated	Moorsh	6 months	10	Chapman and Harris (1981)
	EC, 280 g ha,⁻¹ incorporated	Plainfield sand	2 months	7	Chapman and Harris (1981)
	EC, 140 g ha,⁻¹ not incorporated	Plainfield sand	2 months	4	Chapman and Harris (1981)
Disolfoton (I)	G, 1.12 kg ha⁻¹	Organic soil (82% OMC, pH 5.2, degree of decomposition unspecified)	63 days	<1	Bélanger et Hamilton (1979)
	G, 2.24 kg ha⁻¹	Organic soil (82% OMC, pH 5.2, degree of decomposition unspecified)	63 days	4	Bélanger et Hamilton (1979)
	G, 2.24 kg ha⁻¹	Organic soil (45.4% OMC, pH 5.2, KPI = 20)	30 days	15	Mathur et al. (1980a)

Pesticide	Formulation and rate	Soil	Duration	Value	Reference
Fenpropanate (I)	EC, 280 g ha⁻¹, incorporated	Moorsh	6 months	6	Chapman and Harris (1981)
	EC, 140 g ha⁻¹, not incorporated	Moorsh	6 months	17	Chapman and Harris (1981)
	EC, 280 g ha⁻¹, incorporated	Plainfield sand	2 months	10	Chapman and Harris (1981)
	EC, 140 g ha⁻¹, not incorporated	Plainfield sand	2 months	5	Chapman and Harris (1981)
Fenvalerate (I)	EC, 280 g ha⁻¹, incorporated	Moorsh	6 months	25	Chapman and Harris (1981)
	EC, 140 g ha⁻¹, not incorporated	Moorsh	6 months	15	Chapman and Harris (1981)
	EC, 280 g ha⁻¹, incorporated	Plainfield sand	2 months	22	Chapman and Harris (1981)
	EC, 140 g ha⁻¹, not incorporated	Plainfield sand	2 months	11	Chapman and Harris (1981)
Fonofos (I)	G, 100 mg per 20 cm of row	Peat soil (87% OMC)	119 days	47	Bélanger and Mathur (1983)
	G, 100 mg per 20 cm of row	Moorsh soil (75% OMC)	119 days	45	Bélanger and Mathur (1983)
	G, 1 kg ha⁻¹	Organic soil (KPI = 27)	100 days	21	Bélanger et al. (1982)
	G, 1 kg ha⁻¹	Organic soil (KPI = 90)	100 days	39	Bélanger et al. (1982)
	G, 1 kg ha⁻¹	Organic soil (KPI = 95)	100 days	46	Bélanger et al. (1982)
	G, 1 kg ha⁻¹	Organic soil (KPI = 115)	100 days	48	Bélanger et al. (1982)
Isazophos (I)	EC, 5.94 mg kg⁻¹	Moorsh	24 weeks	7	Chapman and Harris (1982)
	EC, 1.86 mg kg⁻¹	Plainfield sand	24 weeks	1	Chapman and Harris (1982)
Isofenphos (I)	G, 7.44 mg kg⁻¹	Moorsh	2 years	14	Chapman and Harris (1982)
	G, 3.15 mg kg⁻¹	Plainfield sand	2 years	4	Chapman and Harris (1982)
Permethrin (I)	G, 2.24 kg ha⁻¹	Organic soil (45.4% OMC, pH 5.2, PI = 20)	30 days	65	Mathur et al. (1980a)
	G, 0.56 kg ha⁻¹	Organic soil (82% OMC, pH 5.2, degree of decomposition unspecified)	63 days	73	Bélanger and Hamilton (1979)
	G, 1.12 kg ha⁻¹	Organic soil (82% OMC, pH 5.2, degree of decomposition unspecified)	63 days	50	Bélanger and Hamilton (1979)
	EC, 280 g ha⁻¹, incorporated	Moorsh	6 months	1	Chapman and Harris (1981)
	EC, 140 g ha⁻¹, not incorporated	Moorsh	6 months	20	Chapman and Harris (1981)
	EC, 280 g ha⁻¹, incorporated	Plainfield sand	2 months	<1	Chapman and Harris (1981)
	EC, 140 g ha⁻¹, not incorporated	Plainfield sand	2 months	<8	Chapman and Harris (1981)

Note: EC = emulsifiable concentrate; G = granular; S = solution; WP = wettable powder; OMC = organic matter content; KPI = Kaila pyrophosphate index.

amounts of carbofuran residue (> 0.02 mg kg^{-1}). Only 8 of the soils contained more than 0.5 mg kg^{-1} of total carbofuran + 3-ketocarbofuran — highest value was 1.5 mg kg^{-1}. No 3-hydroxycarbofuran was detected. In the same soils, Miles and Harris (1978) and Miles et al. (1978) found residues of DDT (organochlorine) and ethion (organophosphorus) at maximum concentrations of 60 mg kg^{-1} and 25 mg kg,$^{-1}$ respectively. Chapman et al. (1984) found that ethion accumulated more than other insecticides in an organic soil (65% SOM, unspecified decomposition stage) following three consecutive annual applications at the recommended rate under field conditions. Residues in the soil after 3 years were in the order of ethion (7.6 mg kg^{-1}) > chlorpyrifos (1.43 mg kg^{-1}) > chlorfenvinphos (0.58 mg kg^{-1}) > fonofos (0.48 mg kg^{-1}) > carbofuran (< 0.01 mg kg^{-1}). The herbicide nitrofen was found to persist in soil from 6 organic soil farms near Thedford, Ontario, where the highest concentration observed was 35 mg kg^{-1} (Murty et al., 1982). These residues indicated heavy usage and persistence of nitrofen. The SOM degree of decomposition can also affect pesticides persistence, as demonstrated for fonofos (Table 8.3) (Bélanger et al., 1982).

Because of the role of SOM on pesticide retention, one can expect a longer persistence of pesticides in organic rather than in mineral soils. In a survey on insecticides residues in soils of southwestern Ontario, Harris et al. (1977) found that organophosphorus insecticides accumulated significantly more in organic than in mineral soils, with a tenfold increase for residues in organic soils. In a series of experiments conducted in microplots, Chapman and Harris (1980, 1981, 1982) observed that the persistence of organophosphate and pyrethroid insecticides was higher when applied to a moorsh soil compared with a Plainfield sand (Table 8.3). In the same soils, Murty et al. (1982) found that nitrofen persistence was higher in the organic soil. Chapman et al. (1983) observed that fenvalerate accumulated in a moorsh soil after seven applications at 2-week intervals, while its concentration remained relatively constant in a Plainfield sand. Fenvalerate levels declined immediately after spraying in the mineral soil, reaching concentrations of 0.1–0.3 mg kg^{-1} after 2 weeks. In the organic soil, the rate of fenvalerate addition exceeded the rate of disappearance and the concentration in the soil increased to 0.9–1.0 mg kg^{-1} after 14 weeks. The concentration decreased slowly once the applications ceased but was still 0.5–0.7 mg kg^{-1} the following spring. More recently, Szeto and Price (1991) did a survey on pesticides content in the soils located on 12 farms involved in vegetable farming for at least 25 years in the Fraser Valley of British Colombia, Canada. Organochlorine pesticides, which are banned for use in Canada since 1971, were found at higher levels in moorsh than in mineral soils. On the other hand, Szeto et al. (1990) found no difference in the dissipation rate of granular phorate (THIMET 15G) in a silt loam soil and a moorsh soil cultivated with potatoes, with dissipation half-lives of 65 and 64 days, respectively.

V. EFFECTS OF PESTICIDES ON MICROFLORA AND BIOCHEMICAL PROCESSES

The presence of pesticide residues during a prolonged period in organic soils may extend their influence on soil microflora and biochemical processes in organic soils (Table 8.4). Bacteria, actinomycetes, and fungi activities are affected by some pesticides. Such effects are usually temporary, and microbial populations recover

Table 8.4 Effect of Selected Pesticides Applied to Organic Soils on Microbial Populations and Biochemical Processes

Pesticide Common Name	Observed Effects	References
1,3-D (N)	Increase in microbial populations, especially for fungi.	Mathur et al. (1980b)
	Temporary inhibition of the activity of nitrifying bacteria.	Wolcott et al. (1960)
Aldicarb (I)	No effects on microbial population.	Mathur et al. (1980b)
Bromoxynil (H)	Inhibition of methane oxidation at 50 mg L^{-1}, but no effect at 5 mg L^{-1}.	Topp (1993)
Carbofuran (I)	Increase in microbial populations, especially for fungi.	Mathur et al. (1980b)
Disulfoton (I, A)	Temporary effects on microbial population.	Mathur et al. (1980a)
Fenamiphos (N)	Stimulation of microbial populations up to 9 days after application, followed by a suppression of bacteria and actinomycete at 30 days.	Mathur et al. (1980b)
Glufosinate ammonium (H)	Temporary decrease in carboxymethyl cellulase activity in soil.	Ismail and Wong (1994)
Imazapyr (H)	Temporary decrease in carboxymethyl cellulase activity in soil.	Ismail and Wong (1994)
Linuron (H)	Temporary suppression of bacteria and actinomycete; increase in the number of fungal propagules.	Mathur et al. (1976)
Metham sodium (N)	Increase in microbial populations, especially for fungi	Mathur et al. (1980b)
Methomyl (I)	Inhibition of methane oxidation at 50 mg L^{-1}, but no effect at 5 mg L^{-1}.	Topp (1993)
Monolinuron (H)	Stimulation of the growth of fungi after 2 weeks; Inhibition of nitrification after 2 weeks.	Tu (1996)
Nitrapyrin (B)	Inhibition of methane oxidation at 50 mg L^{-1}, but no effect at 5 mg L^{-1}.	Topp (1993)
Oxamyl (I, A, N)	Increase in microbial population, especially for fungi.	Mathur et al. (1980b)
Paraquat (H)	No change in CO_2 evolution; Temporary inhibition of fungi; Increase in bacterial and actinomycetal populations; Increase in NO_3-N levels in treated plots as compared with the control by 74% and 70% at 13 and 33 days after application.	Mathur et al. (1976)
Permethrin (I)	Slight but lasting suppression of microbial population and decreased available N and P in the treated soil as compared with the control.	Mathur et al. (1980a)
Prometryn (H)	Temporary effects on microbial population.	Mathur et al. (1980a)
Simazine (H)	Stimulation of the growth of fungi after 2 weeks; Stimulation of sulfur oxidation after 4 weeks and of denitrification after 2 weeks; Inhibition of nitrification after 2 weeks.	Tu (1996)
Tridiphane (H)	Stimulation of sulfur oxidation after 4 weeks and of denitrification after 2 weeks; inhibition of nitrification after 2 weeks.	Tu (1996)

Note: N = nematicide; H = herbicide; I = insecticide; A = acaricide; B = bactericide.

rapidly (Mathur et al., 1976; Mathur et al., 1980a; Tu, 1979, 1981). Only in a soil where permethrin was added were the effects more important and permanent, which was attributed to the persistence of this insecticide throughout the growing season (Mathur et al., 1980a). Furthermore, the effect of permethrin appeared to be modified by the crop, with a greater suppressive effect on soil populations where carrots were grown compared with lettuce. The presence of bound residues may have an inhibitory effect on respiratory activity of microbes in the soil. Khan and Ivarson (1981) observed that the respiration rate of an organic soil containing ^{14}C-prometryn bound residues was lower compared with control.

Tu (1979, 1981) studied the effect of 32 pesticides on biochemical and microbial activities in an organic soil (27% organic C, pH 7.2, unspecified decomposition stage). None of the pesticides suppressed acetylene reduction compared with control. No significant inhibition of the nonsymbiotic N fixers occurred. However, stimulatory effects were observed with some pesticides. Bacterial and fungal populations showed temporary declines, but all recovered within 7 days to levels similar to or higher than those in the control. Dehydrogenase, phosphatase, and urease activities decreased after the addition of some pesticides, but only temporarily.

Nitrifying bacteria can be inhibited (Wolcott et al., 1960) or stimulated (Mathur et al., 1976) by some pesticides. The significant inhibition of nitrifying bacteria altered the seasonal distribution of ammonium and nitrate in the soil. During the period of retarded nitrification, ammonium N accumulated. This accumulation was probably enhanced in organic soils by partial sterilization of some pesticides, which resulted in a more rapid ammonification of organic N (Wolcott et al., 1960).

Some pesticides showed no effect on microbial populations and biochemical processes. According to Topp (1993), immobilization or degradation cannot be discounted as mechanisms of detoxification. In his study on the effect of 30 different pesticides on methane oxidation in a saprist soil, only three showed an inhibitory effect, which was concentration dependent. The effect of two pesticides, bromoxynil and methomyl, was short-lived, because no significant inhibitory effect on methane oxidation was observed 3 weeks after application. This was attributed to rapid pesticides sorption or degradation in soil. Nitrapyrin, a nitrification inhibitor, inhibited CH_4 oxidation almost completely during 7 weeks. Given similitude between autothrophic nitrifiers and methanotrophs, parallel sensitivity to inhibitors tested in pure culture, and co-metabolism of NH_4^+ or CH_4 by methanotrophs or nitrifiers, respectively, nitrification inhibitors can be expected to inhibit methane oxidation in soil (Bédard and Knowles, 1989).

VI. CONCLUDING REMARKS

The high organic matter content in organic soils is the key factor rendering this environment prime territory for pesticide reactions. The persistence of pesticides in organic soils is highly controlled by their retention, which influences processes such as plant uptake, off-site movements, and degradation. Pesticides applied to organic soils can affect microbial populations and biochemical processes. For most compounds, however, only transient effects were observed. The crop being grown can influence the effect of some pesticides on soil microflora and biochemical processes.

Although the importance of organic matter in the fate of pesticides in the soil environment is well recognized, a relatively small number of studies have been conducted on the fate of pesticides in organic soils. Much work still needs to be done in order to fully understand the long-term impact of the application of pesticides to organic soils. The effect of specific factors such as soil pH, pesticide formulation, soil moisture and temperature on adsorption, degradation and general persistence of pesticides in organic soils needs to be assessed in a rigorous manner, considering pesticide types and soil organic matter composition. The importance of pesticide movement by wind and water erosion, and by colloid-facilitated transport in organic soils, also needs to be studied. Finally, the fate of pesticide-bound residues in organic soils should be assessed to evaluate their potential contribution to pollution and possible effects on plants and soil microorganisms.

REFERENCES

Almendros, G. 1995. Sorptive interactions of pesticides in soils treated with modified humic acids. *Eur. J. Soil Sci.,* 46:287–301.

ARS Pesticide Properties, www.arsusda.gov/acsl/ppdb.html

Bédard, C. and Knowles, R. 1989. Physiology, biochemistry, and specific inhibitors of CH_4, NH_4^+, and CO oxidation by methanotrophs and nitrifiers. *Microbiol. Rev.,* 53:68–84.

Bélanger, A. and Hamilton, H.A. 1979. Determination of disulfoton and permethrin residues in an organic soil and their translocation into lettuce, onion and carrots. *J. Environ. Sci. Health,* B14: 213–226.

Bélanger, A., Mathur, S.P. and Martel, P. 1982. The effect of fonofos and carbofuran on microbial populations, and persistence of fonofos in four soils infested with onion maggot. *J. Environ. Sci. Health,* B17:171–182.

Bélanger, A. and Mathur, S.P. 1983. Persistence of fonofos in two organic soils each containing four levels of copper. *J. Environ. Sci. Health,* B18:713–723.

Bergström, L. 1995. Leaching of dichlorprop and nitrate in structured soil. *Environ. Pollut.,* 87:189–195.

Bollag, T.M. 1992. Biological and chemical interactions of pesticides with soil organic matter. *Sci. Total Environ.,* 123/124:205–217.

Braunschweiler, H. 1992. The fate of some pesticides in Finnish cultivated soils. *Agric. Sci. Finl.,* 1:37–55.

Braverman, M.P. et al. 1990a. Sorption and degradation of thiobencarb in three Florida soils. *Weed Sci.,* 38:583–588.

Braverman, M.P. et al. 1990b. Mobility and bioactivity of thiobencarb. *Weed Sci.,* 38:607–614.

Chapman, R.A. and Chapman, P.C. 1986. Persistence of granular and EC formulations of chlorpyrifos in a mineral and an organic soil in open and closed containers. *J. Environ. Sci. Health,* B21:447–456.

Chapman, R.A. and Harris, C.R. 1980. Persistence of chlorpyrifos in a mineral and an organic soil. *J. Environ. Sci. Health,* B15:39–46.

Chapman, R.A. and Harris, C.R. 1981. Persistence of four pyrethroid insecticides in a mineral and an organic soil. *J. Environ. Sci. Health,* B16:605–615.

Chapman, R.A. and Harris, C.R. 1982. Persistence of isofenphos and isazophos in a mineral and an organic soil. *J. Environ. Sci. Health,* B17:355–361.

Chapman, R.A., Harris, C.R., and Harris, C. 1986. Observations on the effect of soil type, treatment intensity, insecticide formulation, temperature and moisture on the adaptation and subsequent activity of biological agents associated with carbofuran degradation in soils. *J. Environ. Sci. Health,* B21:125–141.

Chapman, R.A. et al. 1984. Persistence and mobility of granular insecticides in an organic soil following furrow application for onion maggot control. *J. Environ. Sci. Health,* B19:259–270.

Chapman, R.A. et al. 1983. Fenvalerate concentrations in a mineral and an organic soil receiving multiple applications during the growing season. *J. Environ. Sci. Health,* B18:685–690.

Chapman, R.A. et al. 1981. Persistence of five pyrethroid insecticides in sterile and natural, mineral and organic soils. *Bull. Environ. Contam. Toxicol.,* 26:513–519.

Chapman, R.A., et al. 1985. Persistence of diflubenzuron and BAY SIR 8514 in natural and sterile sandy loam and organic soils. *J. Environ. Sci. Health,* B20:489–497.

Cheah, U.B., Kirkwood, R.C. and Lum, K.Y. 1998. Degradation of four commonly used pesticides in Malaysian agricultural soils. *J. Agric. Food Chem.,* 46:1217–1223.

Chiou, C.T. 1990. Roles of organic matter, minerals, and moisture in sorption of nonionic compounds and pesticides by soil, *Humic Substances in Soil and Crop Sciences: Selected Readings.* MacCarthy, P. et al., Eds., American Society of Agronomy, Madison, WI, 111–160.

CPVQ, 1997. *Mauvaises Herbes — Répression.* Conseil des Productions Végétales du Québec, Inc. AGDEX 640, Quebec, Canada.

Fadayomi, O. and Warren, G.F. 1977. Adsorption, desorption, and leaching of nitrofen and oxyflurofen. *Weed Sci.,* 25:97–100.

Franco, I. et al. 1997. Adsorption of linuron and metamitron on soil and peat at two different decomposition stages. *J. Soil Contam.,* 6:307–315.

Frank, R., Clegg, B.S. and Ritcey, G. 1991. Disappearance of oxyflurofen (Goal) from onions and organic soils. *Bull. Environ. Contam. Toxicol.,* 46:485–491.

Harris, C.R., Chapman, R.A., and Miles, J.R. 1977. Insecticides residues in soils on fifteen farms in Southwestern Ontario, 1964–1974. *J. Environ. Sci. Health,* B12:163–174.

Hatcher, P.G. et al. 1993. Use of high resolution ^{13}C-NMR to examine the enzymatic covalent binding of ^{13}C-labeled 2,4-dichlorophenol to humic substances. *Environ. Sci. Technol.,* 27:2098–2103.

Hitchings, E.J. and Roberts, T.R. 1980. Degradation of the herbicide flamprop-methyl in soil. *Pestic. Sci.,* 11:591–599.

Hsu, T.S. and Bartha, R. 1974. Biodegradation of chloroaniline-humus complexes in soil and in culture solutions. *Soil Sci.,* 118:213–220.

Ismail, B.S. and Wong, L.K. 1994. Effects of herbicides in cellulotic activity in peat soil. *Microbios.,* 78:117–123.

Jourdan, S.W., Majek, B.A., and Ayeni, A.O. 1998. Imazethapyr bioactivity and movement in soil. *Weed Sci.,* 46:608–613.

Kaila, A. 1956. Determination of the degree of humification in peat samples. *J. Sci. Agric. Soc. Finland,* 28:18–35.

Kästner, M. et al. 1999. Formation of bound residues during microbial degradation of [14C]anthracene in soil. *Appl. Environ. Microbiol.,* 65:1834–1842.

Khan, S.U. 1973. Interaction of humic substances with bipyridillium herbicides. *Can. J. Soil Sci.,* 53:199–204.

Khan, S.U. 1980. Plant uptake of unextractable (bound) residues from an organic soil treated with prometryn. *J. Agric. Food Chem.*, 28:1096–1098.

Khan, S.U. 1982. Distribution and characteristics of bound residues of prometryn in an organic soil. *J. Agric. Food Chem.*, 30:175–179.

Khan, S.U. 1988. Bound residues, in *Environmental Chemistry of Herbicides,* Volume II. Grover, R. and Cessna, A.J., Eds., CRC Press, Boca Raton, FL, 266–279.

Khan, S.U., et al. 1976a. Residues of paraquat and linuron in an organic soil and their uptake by onions, lettuce and carrots. *Can. J. Soil Sci.,* 56:407–412.

Khan, S.U. and Hamilton, H.A. 1980. Extractable and bound (non-extractable) residues of prometryn and its metabolites in an organic soil. *J. Agric. Food Chem.,* 28:126–132.

Khan, S.U., Hamilton, H.A., and E.J. Hogue. 1976b. Fonofos residues in an organic soil and vegetable crops following treatment of the crop with the insecticide. *Pestic. Sci.,* 7:553–558.

Khan, S.U. and Ivarson, K.C. 1981. Microbial release of unextracted (bound) residues from an organic soil treated with prometryn. *J. Agric. Food Chem.,* 29:1301–1303.

Kile, D.E. et al. 1995. Partition of non-polar organic pollutants from water to soil and sediment organic matters. *Environ. Sci. Technol.,* 29:1401–1406.

Larney, F.J., Cessna, A.J., and Bullock, M.S. 1999. Herbicide transport on wind-eroded sediment. *J. Environ. Qual.,* 28:1412–1421.

Leistra, M. and Green, R.E. 1990. Efficacy of soil-applied pesticides, in *Pesticides in the Soil Environment: Processes, Impacts, and Modeling.* Cheng, H.H., Ed., Soil Science Society of America Book Series, Number 2, Madison, WI, 401–428.

Leonard, R.A. 1990. Movement of pesticides into surface waters, in *Pesticides in the Soil Environment: Processes, Impacts, and Modeling.* Cheng, H.H., Ed., Soil Science Society of America Book Series, Number 2, Madison, WI, 303–400.

Mathur, S.P. et al. 1980a. Influence on soil microflora and persistence of field-applied disulfoton, permethrin and prometryn in an organic soil. *Pedobiologia.,* 20:237–242.

Mathur, S.P. et al. 1976. Influence of filed-applied linuron and paraquat on the microflora of an organic soil. *Weed Research.,* 16:183–189.

Mathur, S.P. and Farnham, R.S. 1985. Geochemistry of humic substances in natural and cultivated peatlands, in *Humic Substances in Soil, Sediment, and Water. Geochemistry, Isolation and Characterization.* Aiken, G.R. et al., Eds., John Wiley & Sons, New York, 53–85

Mathur, S.P., Hamilton, H.A., and Vrain, T.C. 1980b. Influence of some field-applied nematicides on microflora and mineral nutrients in an organic soil. *J. Environ. Sci. Health.,* B15:61–76.

Mathur, S.P. and Morley, H.V. 1975. A biodegradation approach for investigating pesticide incorporation into soil humus. *Soil Sci.,* 120:238–240.

Mathur, S.P. and Morley, H.V. 1978. Incorporation of methoxychlor-^{14}C in model humic acids prepared from hydroquinone. *Bull. Environ. Contam. Toxicol.,* 20:268–274.

Miles, J.R.W. and Harris, C.R. 1978. Insecticide residues in water, sediment, and fish of the drainage system of the Holland marsh, Ontario, Canada, 1972–1975. *J. Econ. Entomol.,* 71:125–131.

Miles, J.R.W. and Harris, C.R. 1979. Carbofuran residues in organic soils in southwestern Ontario. *J. Environ. Sci. Health,* B14:655–661.

Miles, J.R.W., Harris, C.R., and O. Moy. 1978. Insecticide residues in organic soil of the Holland marsh, Ontario, Canada, 1972–1975. *J. Econ. Entomol.,* 71:97–101.

Miles, J.R.W., Harris, C.R., and Tu, C.M. 1984. Influence of moisture on the persistence of chlorpyrifos and chlorfenvinphos in sterile and natural mineral soils. *J. Environ. Sci. Health,* B19:237–243.

Miles, J.R.W., Tu, C.M., and Harris, C.R. 1981. A laboratory study of the persistence of carbofuran and its 3-hydroxy- and 3 keto-metabolites in sterile and natural mineral and organic soils. *J. Environ. Sci. Health,* B16:409–417.

Morita, H. 1976. Linuron adsorption and the degree of decomposition of peats as measured by rubbed fiber content and pyrophosphate index. *Can. J. Soil Sci.,* 56: 105–109.

Morris, R.F. and Penny, B.G. 1971. Persistence of linuron residues in soils at Kelligrews, Newfoundland. *Can. J. Plant Sci.,* 51:242–245.

Murty, A.S., Miles, J.R.W., and Tu, C.M. 1982. Persistence and mobility of nitrofen (Noclofen, TOF^R) in mineral and organic soils. *J. Environ. Sci. Health,* B17:143–152.

Parent, L.E. and Bélanger, A. 1985. Comparison between Freundlich and fixed regression models of linuron retention by soil organic materials. *J. Environ. Sci. Health,* A20:293–304.

Pennington, K.L., Harper, S.S., and Koskinen, W.C. 1991. Interactions of herbicides with water-soluble soil organic matter. *Weed Sci.,* 39:667–672.

Raman, S. and Rao, C. 1988. The kinetics of extraction of soil-applied metoxuron by methanol and its biological implications. *Water Air Soil Pollut.,* 38:217–222.

Rutherford, D.W., Chiou, C.T., and Kile, D.E. 1992. Influence of soil organic matter composition on the partition of organic compounds. *Environ. Sci. Technol.,* 26:336–340.

Sahid, I.B. and Teoh, S.S. 1994. Persistence of terbutylazine in soils. *Bull. Environ. Contam. Toxicol.,* 52:226–230.

Schwarzenbach, R.P., Gschwend, P.M., and Imboden, D.M. 1993. *Environmental Organic Chemistry.* John Wiley & Sons, New York,

Scott, D.E. et al. 1983. Biodegradation, stabilization in humus, and incorporation into soil biomass of 2,4-D and chlorocatechol carbons. *Soil Sci. Soc. Am. J.,* 47:66–70.

Senesi, N. 1992. Binding mechanisms of pesticides to soil humic substances. *Sci. Total Environ.,* 123/124:63–76.

Senesi, N. and Chen, Y. 1989. Interactions of toxic organic chemicals with humic substances, in *Toxic Organic Chemicals in Porous Media. Ecological Studies, Vol. 73.* Gerstl, Z. et al., Eds., Springer-Verlag, Berlin, 37–90.

Senesi, N. and Miano, T.M. 1995. The role of abiotic interactions with humic substances on the environmental impact of organic pollutants in *Environment Impact of Soil Component Interactions. Natural and Anthropogenic Organics, Vol. I.* Huang, P.M. et al., Eds., CRC Press, Boca Raton, FL, 311–335.

Soil Classification Working Group. 1998. Agriculture and Agri-Food Canada Publication 1646 (Revised), 187 pp.

Soil Survey Staff. 1990. *Keys to Soil Taxonomy,* 4th ed. SMSS technical monograph no. 19. Blacksburg, Virginia.

Sterling, T.M. 1994. Mechanisms of herbicide absorption across plant membranes and accumulation in plant cells. *Weed Sci.,* 42:263–276.

Stevenson, F.J. 1976. Organic matter reactions involving pesticides in soil. *ACS Symposium Series,* 29:180–207.

Stevenson, F.J. 1985. Geochemistry of soil humic substances, in *Humic Substances in Soil, Sediment, and Water. Geochemistry, Isolation, and Characterization.* Aiken, G.R. et al., Eds., Wiley Interscience, New York, 13–52.

Stevenson, F.J. 1994. *Humus Chemistry. Genesis, Composition, Reactions,* 2nd ed. John Wiley & Sons, New York, 496 pp.

Suett, D.L. 1975. Persistence and degradation of chlorfenvinphos, chlormephos, disulfoton, phorate and pirimiphosethyl following spring and late-summer soil application. *Pestic. Sci.* 6:385–393.

Szeto, S.Y. and Price, P.M. 1991. Persistence of pesticide residues in mineral and organic soils in the Fraser Valley of British Colombia. *J. Agric. Food Chem.,* 39:1679–1684.

Szeto, S.Y. et al. 1990. Persistence and uptake of phorate in mineral and organic soils. *J. Agric. Food Chem.,* 38:501–504.

Topp, E. 1993. Effects of selected agrochemicals on methane oxidation by an organic agricultural soil. *Can. J. Soil Sci.,* 73:287–291.

Torrents, A., Jayasundera, S., and Schmidt, W.J. 1997. Influence of the polarity of organic matter on the sorption of acetamide pesticides. *J. Agric. Food Chem.,* 45:3320–3325.

Triegel, E.K. and Guo, L. 1994. Overview of the fate of pesticides in the environment, water balance; runoff vs. leaching, in *Mechanisms of Pesticide Movement into Groundwater.* Honeycutt, R.C. and Schabacker, D.J., Eds., Lewis Publishers, Boca Raton, FL, 1–13.

Tu, C.M. 1979. Influence of pesticides on acetylene reduction and growth of microorganisms in an organic soil. *J. Environ. Sci. Health,* B14:617–624.

Tu, C.M. 1981. Effects of some pesticides on enzyme activities in an organic soil. *Bull. Environ. Contam. Toxicol.,* 27:109–114.

Tu, C.M. 1996. Effect of selected herbicides on activities of microorganisms in soils. *J. Environ. Sci. Health,* B31:1201–1214.

Turco, R.F. and Kladivko, E.J. 1994. Studies on pesticide mobility: Laboratory vs. field, in *Mechanisms of Pesticide Movement into Groundwater.* Honeycutt, R.C. and Schabacker, D.J., Eds., Lewis Publishers, Boca Raton, FL, 63–80.

Vaillancourt, N. et al. 1999. Sorption of ammonia and release of humic substances as related to selected peat properties. *Can. J. Soil Sci.,* 79:311–315.

Wall, G.J. et al. 1995. *Erosion. The Health of Our Soils. Toward Sustainable Agriculture in Canada.* Acton, D.F. and Gregorich, L.J., Eds., Centre for Land and Biological Resources Research, Agriculture and Agri-Food Canada Publication 1906/E, Ottawa, Ontario.

Weber, J. B. 1994. Properties and behavior of pesticides in soil, in *Mechanisms of Pesticide Movement into Groundwater.* Honeycutt, R.C. and Schabacker, D.J., Eds., Lewis Publishers, Boca Raton, FL, 15–42.

Wolcott, A.R. et al. 1960. Effects of TELONE on nitrogen transformations and on growth of celery in organic soil. Down to Earth. Summer 1960: 1–5, Journal Article 2595, Dept. Soil Sci., Michigan State University, East Lansing, MI.

Zhang, L.Z. et al. 1984. Persistence, degradation, and distribution of deltametrhin in an organic soil under laboratory conditions. *J. Agric. Food Chem.,* 32:1207–1211.

Table 8.5   Chemical Indexes

Common Name	Chemical Name
1,3-D	1,3-Dichloropropane
2,4-D	2,4-Dichlorophenoxyacetic acid
Alachlor	2-Chloro-2.6'-diethyl-N-[methoxymethyl]acetanilide
Aldicarb	2-Methyl-2-(methylthio)propionaldehyde o-(methylcarbamoyl)oxime
Bromoxynil	3,5-Dibromo-4-hydroxybenzonitrile
Carbofuran	2,3-Dihydro-2.2-dimethyl-benzofuran-7-yl methylcarbamate
Chlorfenvinphos	2,4-Dichloro-alpha-(chloromethylene)benzyl alcohol diethyl phosphate
Chlorpyrifos	O.O-Diethyl-O-[3.5.6-trichloro-2-pyridyl]-phosphorothioate
Cypermethrin	(+/-)(a)-Cyano-3-phenoxybenzyl (+) cis-trans-3-(2.2-dichlorovinyl)-2.2-dimethyl-cyclopropane carboxylate
DDT	1.1.1-Trichloro-2.2-di(4-chlorophenyl)ethane
Deltamethrin	(S)-a-Cyano-3-phenoxybenzyl-(1R,3R)-3-(2,2-dibromovinyl)-2,2-dimethylpropane-carboxylate
Dichlorprop	2-(2.4-Dichlorophenoxy)propionic acid
Diquat	6,7-dihydrodipyrido(1,2–1:2',1'-c)pyradizium salt
Disulfoton	O.O-Diethyl-S-[2-(ethylthio)ethyl]-phosphorodithioate
Ethion	O,O,O',O'-Tetraethyl S,S'-methylene bis(phosphorodithioate)
Fenamiphos	Ethyl 4-(methylthio)-m-totyl isoprophylphosphoramidate
Fenvalerate	(S)-Alpha-cyano-3-phenoxybenzyl(S)-2-(-4-chlorophenyl)-3-methylbutyrate
Flamprop-methyl	N-benzoyl-N-(3-chloro-4-fluorophenyl)-DL-alanine
Fonofos	O-ethyl-s-phenyl ethyl phosphonodithiote
Glufosinate ammonium	Ammonium-DL-homoalanin-4-yl(methyl)-phosphinic acid
Glyphosate	N-(Phosphonomethyl) glycine
Imazapyr	2-(4-Isopropyl-4-methyl-5-oxo-2-imidazolin-2-yl)-nicotinic acid
Isazophos	O.O-Diethyl-O-[1-isopropyl-5-chloro-1,2,4-triazolyl-[3]]-phosphorothioate
Isofenphos	1-Methylethyl-2-[[ethoxy[(1-methylethyl)-amino]phosphinothioyl]-oxy]benzoate
Linuron	3-(3,4-dichlorophenyl)-1-methoxy-1-methylurea
Metamitron	4-Amino-4,5-dihydro-3-methyl-6-phenyl-1,2,4-triazin-5-one
Metham sodium	Sodium methyl dithiocarbamate
Methomyl	S-Methyl-N-[(methylcarbamoyl)oxy]thioacetimidate
Metolachlor	2-Chloro-N-[2-ethyl-6-methylphenyl]-N-[2-methoxy-1-methylethyl] acetamide
Monolinuron	3-(4-Chlorophenyl)-1-methoxy-1-methylurea
Nitrapyrin	2-Chloro-6-(trichloromethyl)pyridine
Nitrofen	2,4-Dichlorophenyl 4-nitrophenyl ester
Oxamyl	Methyl-N',N-dimethyl-N-[(methylcarbamoyl)oxy]-1-thiooamimidate
Oxyfluorfen	2-Chloro-1-(3-ethoxy-4-nitrophenoxy)-4-(trifluoromethyl)benzene
Paraquat	1,1'-dimethyl-4,4' dipyridium salt
Permethrin	[3-Phenoxyphenyl] methyl-[+]-cis-trans-3-[2.2-dichloroethenyl]-2.2-dimethylcycopropane carboxylate
Phorate	O.O-Diethyl-S-[(ethylthio)methyl]phosphorodithioate
Prometryn	2.4-Bis(isopropylamino)-6-methylthio-s-triazine
Propachlor	2'-Chloro-N-isopropyl acetanilide
Sethoxydim	(±)-(EZ)-2-(1-Ethoxyiminobutyl)-5-[2-(ethylthio)propyl]-3-hydroxycyclohex-2-enone
Simazine	2-Chloro-4,6-bis(ethylamino)-s-triazine
Terbuthylazine	2-tert-Butylamino-4-chloro-6-ethylamino-1.3.5-triazine
Thiobencarb	S-(4-Chlorophenyl)methyl diethylcarbamothioate
Tridiphane	(RS)-2-(3,5-Dichlorophenyl)-2-(2,2,2-trichloroethyl)oxirane
Trifluralin	a.a.a-Trifluoro-2.6-dinitro-N.N-dipropyl-p-toluidine

CHAPTER 9

Quality of Organic Soils for Agricultural Use of Cutover Peatlands in Russia

Vera N. Kreshtapova, Rudolf A. Krupnov, and Olga N. Uspenskaya

CONTENTS

ABSTRACT

The area of cutover peatlands in Russia is estimated at 2 million ha. The aim of this chapter is to present organic soil quality criteria for agricultural production on cutover peatlands in Russia. From the analyses of 10,000 soil profiles in European Russia, the authors developed the following soil criteria with the ascending order of capability in parentheses:

- Thickness of the arable peat layer (<15 cm to 30–40 cm)
- Degree of peat decomposition (<20% to >50%)
- C/N ratio (>25 to 10–14)
- Ash content (<0.10 to > 0.40 kg kg^{-1})
- Bulk density (<0.20 to > 0.40 g cm^{-3})
- pH$_{KCl}$ (<4.5 to >6.0)
- Cationic base saturation of the cation exchange capacity (< 20% to > 60%)

Rock outcrops, uneven thickness of the residual peat layers, the presence of a gley layer below peat, toxic compounds, as well as abrupt temperature and moisture fluctuations during the growing season, may reduce the quality of organic soils in reclaimed cutover peatlands.

I. INTRODUCTION

In a natural raised bog, the actotelm (living moss layer) and the catotelm (bulk of the peat deposit) have functional roles in peat accumulation as well as water storage and transmission. No acrotelm is present during peat excavation, and artificial drainage is required. In Russia, commercial peatlands are segmented by main ditches that are 500 m apart. Perpendicular ditches, which are 20 m apart on bogs and 40 m apart on fens, divide the field into rectangular plots. After peat excavation, the cutover peatland, where 25–40 cm of residual peat is left over mineral or limnic materials, often remains as bare soil for 1 to 2 years. Such cutover peatlands and shallow adjacent organic soils cover nearly 2 million ha in Central Russia. Cutover peatlands have been regarded as infertile soil for a long time. In the Moscow region, however, cutover peatlands cover 100,000 ha, of which areas totaling 15,000 ha are now used in agriculture.

The agricultural use of cutover peatlands was supported by successful peat cultivation methods in The Netherlands, Germany, and other countries. For Russian cutover peatlands, current plans are to rebuild the drainage network, deepen ditches to 1.2 m, cut and remove woody vegetation, uproot and remove woody inclusions, level the surface, fill in depressions and small ditches, and prepare the soil before seeding with proper tillage, liming, and fertilization methods (Krupnov and Popov, 1995).

For sustainable use, shallow organic soils should be cultivated or afforested, depending on biophysical and economic considerations (Fatchikhina, 1960; Trutnev, 1963; Korenova, 1982; Krupnov, 1978; Vertogradskaya et al., 1992; Kreshtapova, 1993). In Ireland, future use of cutover peatlands as a percentage of the area was predicted to be softwood forestry (40–50%), hardwood forestry (10–20%), grassland (20–30%), as well as wetlands and natural landscape (20–30%) (McNally, 1995).

Agricultural utilization requires a reliable classification system including chemical properties, especially the presence of toxic substances, properties of the mineral substratum, and initial fertility level. The aim of this chapter is to document the quality of organic soils for agricultural reclamation of cutover peatlands in Russia.

II. MATERIALS AND METHODS

A. Soil Analyses

The authors examined 34 deposits of cutover peatlands and 1000 soil profiles in European Russia during the 1978–1988 and 1990–1997 periods. More than 10,000 samples were taken from genetic horizons and analyzed for botanical composition, degree of decomposition, C/N ratio, ash content, pH, moisture content, bulk density, as well as total and available forms of elements.

Botanical composition was determined as averaged percentages of humus-washed plant debris in 10 samples under the microscope at magnifications of 56× to 140×. The degree of peat decomposition was assessed by sieving and centrifuging peat materials, then determining degree of decomposition on a nomograph, as reported by Parent and Caron (1993). Moisture content was determined by drying 100 to 200 g of chopped (<3–4 mm) peat samples at 105–110°C, and computed as the weight percentage of water relative to the weight of the field-moist peat sample. Bulk density was determined using 50 ml cylinders.

Chemical analyses were carried out according to Sapek and Sapek (1988) and Rin'kis and Nollendorf (1982). Ash content was obtained by burning 6–8 g of chopped (<3–4 mm) and 105–110°C dried peat samples in a muffle furnace at 800 ± 25°C, and computed as weight percentage of loss on ignition divided by the dry peat weight. Ash composition was analyzed from 2 g of ashes obtained after burning the peat at 400°C, then dissolving ashes in a 6 M HCl solution. The authors determined Fe, Al, Ca, and Mg by compleximetric methods (e.g., Na_2EDTA). The K was determined by flame photometry, and P by the phospho-molybdate method. Carbon content was determined as CO_2 adsorbed on ignited lime. Soil pH was obtained after mixing 20 to 30 ml of field-moist peat material with 1 M KCl (pH 5.6–6.0) in a 1:2.5 (v/v) ratio, equilibrating for 18–20 h, and taking final pH of the suspension. Exchangeable hydrogen in 2 g dry peat samples was displaced by 600 ml of CH_3COONa 1 M. A 100-ml aliquot was back-titrated to pH 7.0 using NaOH 0.1 M.

Plant-available nutrients were analyzed using a 1:10 w/v soil to solution ratio of field-moist peat materials. Available P and Fe, and exchangeable cations were extracted using 250 ml of HCl 0.2 M. The P was determined by the phospho-molybdate method using ascorbic acid as a reducing agent, the K by flame photometry, and the Fe, Ca, and Mg by compleximetric (thiocyanate for Fe) or titrimetric (Ca and Mg) methods. Ammonium N was extracted with 250 ml of HCl 0.1 M and determined using the Nessler reaction. Nitrate N was extracted either with 250 ml of distilled water (colorimetric method using phenoldisulfonic acid as reducing agent), or Al-K sulfates (potentiometric method using the nitrate-specific electrode).

B. The Ozeretsko–Nikolskoe Cutover Peatland

The Ozeretsko–Nikolskoe cutover bog occupies the second terrace of the Klyazma River in the Moscow region, Russia. The authors examined plot no. 44 (180 ha), where oat yield (*Avena sativa* L., cv. "Early Ripening"), measured by hand-harvesting on 225 m², plots varied between 0.34 and 7.21 Mg ha⁻¹ (Korenova,

1982). Uneven soil thawing and temperature (2–5°C difference) related to deep or thin peat layers and transitional or fen peat types occurred, overlying sand. Samples were collected in 22 soil profiles. Genetic horizons were sampled in 18 profiles. Four profiles were sampled intensively at least every 20 cm across thick horizons.

C. Statistical Analyses

Statistical analyses comprised basic statistics such as mean and standard deviation, correlation coefficients, and regression analysis.

II. RESULTS AND DISCUSSION

A. Pedological Features of the Surveyed Cutover Peatlands

In Russia, cutover peatlands cover morainic or glaciofluvial plains and terraces. Peatlands laid on flat watersheds without catchment areas (type 1) are most often underlain by clay and loam (subtype 1A), thus maintaining moist surface and high groundwater conditions favorable to sedge, sedge-*Hypnum* and *Hypnum,* as well as fen peat formation. Type 2 peatlands occur in potholes and zones of confluence of groundwater flow, and are colonized by transitional and fen mosses and moss-sedge vegetation. Type 3 peatlands develop on undulating plains of frontal moraine, and consist of alder and, more rarely, *Hypnum* vegetation. Type 4 peatlands occupy zones of transitional groundwater flow, and are made of woody, marsh, sedge, and sedge-*Hypnum* peats, often of high ash content.

In Russia, the soil profiles of cutover peatlands comprise the following horizons:

1. Organic horizons (ash content varying from 3 to 48%) are dominated by fibric to hemic peat of thickness ranging from several centimeters to 1–1.5 m. Underlying horizons made of hemic to sapric peat materials are either absent or thin (3–5 to 25–35 cm). Cutover peatlands do not have a "contact horizon" when underlain by sapropel (limnic materials).
2. Mineral horizons begin with a "contact horizon" made of mineral materials rich in humus, sometimes with inclusions of hemic peat material. The G horizon may be absent in coarse-textured subsoils. Mineral horizons usually show fine texture, high viscosity, and green-blue-gray inclusions. Soil reaction varies from 3.5 to 6.7, and base saturation from 65 to 90%. An illuvial B horizon, made of sand and some podzolized clay, may be found under the contact horizon with the upper peat material. The B horizon is underlain by blue-gray sand, sometimes with inclusions of pebbles, cobbles, or whitish-blue-gray compact clay.

Hydraulic conductivity varied in the range of 0.06–5.2 m d^{-1} in peat to 0.0001–0.09 m d^{-1} in the contact horizon (low values due to the mixture of sapric peat materials and silt), and 0.001–9.5 m d^{-1} in the mineral subsoil. The dense and viscous contact horizon typically showed a massive structure, thus preventing water infiltration and restricting root penetration below the arable layer. Drying of the contact horizon must restrict plant water uptake to the peat layer. During snowmelt and high rainfall, a

Table 9.1   Classification of Cutover Peatlands in Russia

Type of Cutover Peatland after Position in the Landscape	Underlying conditions		
	Impermeable (Clay and Loam) A	Permeable (Sand and Coarse Loamy Sand) B	Pressure Aquifer C
In watershed, no catchment area (1)	1A	1B	1C
In watershed depressions (2)	2A	2B	2C
Near or on river terraces (3)	3A	3B	3C
In floodplain (4)	4A	4B	4C

Source: From Kreshtapova, V.N. 1996. Physical, chemical and geochemical properties of cutover peat areas cultivated in agriculture within the Moscow region, in *Proc. 10th Int. Peat Congr.*, 2:174–183, Bremen, Germany. With permission.

temporary perched water table could occur. Bulk density values ranged from 0.08 to 0.39 g cm^{-3} for upper peat materials to 0.48–1.50 in the contact horizon, and 1.20–1.84 in the mineral subsoil. Particle density varied from 1.40 to 1.65 g cm^{-3} for peat materials, depending on the degree of decomposition and ash content, to 1.60–2.10 and 2.45–2.65 g cm^{-3} in the contact and mineral horizons, respectively.

B. Agricultural Capability of the Surveyed Cutover Peatlands

Reclamation is not recommended for organic soils connected to aquifers (class C), which are suited for establishing wetlands, natural landscapes, and recreational areas. Cutover peatlands of types 1A and 1B (Table 9.1) should be afforested, while those of types 2A and 2B are suited for afforestation or agriculture. Cutover peatlands of types 1A, 1B, 2A, and 2B (Table 9.1) have low soil quality due to subsoil made of blue clay of lacustrine origin, peat pH (KCl 1 M) less than 4.0, and high content in 0.2 M HCl-extractable Fe (up to 3.5%) and Al (up to 1.1%) in peat layers. Organic soils of type 3A, 3B, 4A, and 4B are reserves of agricultural land near large cities. They show high base saturation (60–70%, sometimes 75–80%), optimum acidity (pH of 4.5–6.0 in KCl 1 M), and up to 18 Mg N ha^{-1} in the 0–25 cm layer.

Compared to the bottom peat layer of a natural peatland, residual peat materials in Russian cutover peatlands showed higher ash content, lower moisture content, higher degree of decomposition, and higher nutrient content (Table 9.2). Fen peats generally showed higher pH and base saturation than bog peats, and thus required less lime before cultivation (Table 9.3). In milled-peat extraction fields recently abandoned in European Russia, soil fertility is low. The peat-forming process may start over again. Grass (e.g., *Calamagrostis* sp., *Comarum palustre*, *Deschampsia caespitosa*, *Nadrue stricta*, *Festuca ovina*, *Molinia* sp.) coverage may reach 30–40% after 3 years of abandonment if the groundwater table (GWT) is at least 0.9 m below surface (Table 9.4). With higher GWT, sedges (e.g., *Carex rostrate*, *Carex stricta*, *Carex paradoxa*, *Carex lasiocarpa*) rapidly colonize the abandoned peatland. Composition and productivity of natural vegetation depend on mainly GWT, as follows:

1. Among wood species, birch trees (*Betula pubescens*, *Betula humilis*) growing at a rate of 0.13 to 3.22 Mg ha^{-1} yr^{-1} dominate (50–100%) in well-drained areas,

Table 9.2   Mean Characteristics of Peat near the Bottom of Peatlands before Excavation and of the Peat Layer of Cutover Peatlands Underlain by Mineral Substratum in European Russia

	Substratum[a]				
Parameters	Sand	Loamy Sand	Loam	Clay	Sapropel
pH Unit					
pH$_{KCl}$	5.4	5.2	4.6	4.4	5.5
	4.6	5.4	5.0	4.6	5.8
g kg^{-1} on a Field-Moist Basis					
Moisture content (105°C)	855	860	867	878	886
	753	760	810	821	856
%					
Degree of decomposition	34.9	34.5	34.5	34.7	31.0
(%) (Russian method)	41.9	38.6	38.0	38.0	33.4
g kg^{-1} on a Dry-Weight Basis					
Ash content (800°C)	169	142	96	70	111
	177	151	141	95	122
Kjeldahl-N	17.9	1.8.4	1.6.9	16.8	18.0
	18.1	1.8.5	1.7.1	17.0	18.3
Ca	21.6	23.3	16.6	15.2	22.9
	23.4	23.3	21.7	17.9	24.1
Fe	9.4	8.4	6.6	3.2	9.1
	17.3	17.0	16.2	12.6	12.6
mg kg^{-1}					
P	611	480	480	437	393
	830	611	655	830	961
K	2250	2083	2083	1667	1167
	2333	2250	2250	2167	1580

[a] Above the line are peat characteristics at the bottom of peatland before mining; below the line, after peat excavation.

Table 9.3   Chemical Properties of Fen and Bog Peat Materials Making up the Residual Peat Layer in Cutover Peatlands of European Russia

	Fen Peat[a]			Bog Peat[a]		
Characteristics	W	W-S	S	W[a]	W-S	S
Decomposition degree (%)	58.0	58.0	35.0	50.0	55.0	45.0
Ash content (g kg^{-1})	140.0	132.0	84.0	82.0	166.0	37.0
pH$_{KCl}$	5.9	4.8	4.5	3.0	3.8	4.2
Exchangeable acidity (cmol$_c$ kg^{-1})	35.0	64.2	48.7	126.4	107.4	59.0
Cation exchange capacity (cmol$_c$ kg^{-1})	190.8	102.0	83.4	34.8	47.7	40.6
Cationic base saturation (cmol$_c$ kg^{-1} per cmol$_c$ kg^{-1})	0.845	0.614	0.631	0.216	0.308	0.408

[a] W = woody; S = sedge.

Table 9.4   Species Composition as % Coverage in 20 m × 20 m Quadrants as a Function of Groundwater Level (GWL), 3 Years after Abandonment of Cutover Peat-Milling Fields

GWL (m)	Surface Coverage %[a]			
	Herbaceous Vegetation		Woody–Shrubby Vegetation	
	Sedges[a]	Grasses[a]	Birch[a]	Willow[a]
0.2	67 50–100	0 0	24 0–30	72 50–100
0.4	50 30–100	6 0–20	50 0–90	34 0–10
0.6	31 0–75	12 0–30	54 30–80	31 10–60
0.9	18 0–50	33 0–60	66 50–90	20 0–40
1.2	16 0–50	38 0–50	72 30–95	22 0–60
1.4	15 0–30	42 30–75	97 95–100	3 0–5

[a] Numerators are mean values; denominators are minimum and maximum values.

and willows (*Salix myrtilloides, Salix cineree, Salix lapponum*) dominate in poorly drained areas.

2. Sedges (*Carex lasiocarpa, Carex rostrata, Carex stricta, Carex diandre*) (30–100%) dominate when groundwater level is within 0.4 m below surface, but willows may also cover 50–100% of the ground.

3. In patches of temporary stagnant water, horsetail (*Equisetum limosum*) and *Sphagnum (S. obtusum, S. subsecundum, S. warnstorfii. S. magellanicum*) occur throughout sedge communities. With continuous stagnation, celandine (*Chelidonium generis, Chelidonium majus*), reed (*Phragmites communis*), and cattail (*Typha latifolia*) predominate.

C. Soil Potential of the Ozeretsko-Nikolskoe Type 3A Cutover Peatland in the Moscow Region

In the Ozeretsko–Nikolskoe cutover peatland, peat thickness ranged from 0 to 50 cm along 1-m intervals of a 70-m transect. Chemical properties varied as follows: 280 to 17, 990 mg Kjeldahl N kg^{-1}, 65 to 808 mg total $P kg^{-1}$, 48 to 262 mg available $P kg^{-1}$, 1104 to 3958 mg total $K kg^{-1}$, 7 to 233 mg exchangeable $K kg^{-1}$, and pH_{KCl} of 3.7 to 4.5. At field scale, irregularly distributed 1- to 15-m^2 areas produced varying patterns of crop yields.

Two ecological units were present. Crops failure was associated with transitional bogs with thick peat overlying an impermeable contact horizon. According to Trutnev (1963), soil atmosphere must be enriched with CO_2 (up to 12% v/v) while O_2 is kept low (up to 2% v/v) in such type A soils. As a result, roots remained at soil surface, especially during wet seasons. Ecological units in elevated parts of the microrelief, where oats yielded 1.5 to 2.0 Mg ha^{-1} or more, were transitional and fen organic soils were made of thin peat overlying sand. There, N, P, K, and Al contents in the arable layer, as well as thickness of the arable layer, were not related to crop yield (Table 9.5); however, cationic base saturation and ash content were higher in the arable layer of the most productive sites.

In peat layers, porosity varied between 0.89 and 0.93 m^3 m^{-3}, and bulk density between 0.11 and 0.19 g cm^{-3}. Particle density was 1.6 g cm^{-3}. Water deficit varied between 46 and 165 mm during the growing season. A 10 to 40 cm peat layer overlying sand may be disconnected from groundwater. Maximum water capillary rise through peat was 37.7 cm after 40 days of equilibration. In the contact horizon, total porosity varied between 0.33 and 0.78 m^3 m^{-3}, bulk density between 0.46 and 1.28 g cm^{-3}, and particle density between 0.93 and 2.30 g cm^{-3}. The cultivated cutover sedge-*Hypnum* peat may take 6 years to increase its bulk density from 0.14–16 to 0.22–0.34 g cm^{-3}. When the contact horizon becomes dry, changes in physical properties are irreversible. Soil moisture may reach wilting point, thus leading to crop failure. After 6 years of agricultural use, moisture content in the soil decreased by 219%, and drought became a crop limiting factor in the Ozeretsko–Nikolskoe cutover peatland.

D. Quality Indicators of Reclaimed Cutover Peatlands

According to Fatchikhina (1960) and Trutnev (1963), productive cutover peatlands showed an arable layer exceeding 40 cm and made of sapric peat materials with an ash content more than 0.40 kg/kg^{-1} and a bulk density exceeding 0.4 g cm^{-3}. Chemical properties were as follows: pH_{KCl} not less than 6, cationic base saturation more than 75%, average 0.2 M HCl-extractable levels of 80 mg Fe/kg^{-1}, 114 mg P kg^{-1}, and 483 mg K kg^{-1}, and C/N ratios exceeding 10–14. Some Fe compounds may be phytotoxic or may decrease P availability. Because soils of cutover peatlands have low biological activity, NO_3-N must limit crop growth; however, NO_3-N has high mobility and variability. Therefore, NH_4-N is a preferable indicator of the N status. The authors thus propose the following indicators to monitor organic soil quality on cutover peatlands: thickness of the plow layer and residual peat; degree of decomposition; ash content; bulk density; pH; base saturation; content of 0.2 M HCl-extractable Fe, P, and K; content of NH_4-N; and C/N ratio (Table 9.6). Nutrient content should consider bulk density when expressed in kg/ha^{-1}.

At the start of reclamation, soils of cutover peatlands show neither an acrotelm nor an arable layer. Thickness of the arable layer should exceed 40 cm for productive soils (Table 9.6). Because organic soils subside at a rate of 1–1.5 cm yr^{-1}, peat thickness should be not less than 30–40 cm for reclamation. Five to 8 years after peat extraction, a root layer up to 10–15 cm thick is formed. Highest yields on cutover peatlands are obtained when residual peat is 50 to 100 cm thick. The C/N ratio is used to evaluate the quality of the humus produced by the biochemical transformation of organic soils. The C/N ratio decreases from 20–27 in fen peat to 10–14 in fertile soils. Ash content increases with fertilization and liming, mineral inputs from groundwater and dust, as well as additions of mineral subsoil during plowing and ditching operations. Ash content reaches 0.40–0.50 kg kg^{-1} in fertile soil. Bulk density of productive soils is 0.40–0.60 g cm^{-3}.

Table 9.5 Chemical Characteristics of Soil Profiles in Nonproductive and Productive Cereal Crops Grown on the Ozeretsko–Nikolskoe Cutover Peatland in the Moscow Region

Horizon	Depth (cm)	Ash (g kg⁻¹)	pH_{KCl}	Al^a	$Acidity^b$ (cmol$_c$ kg⁻¹)	$\Sigma\ Bases^c$	CBS^d	N (g kg⁻¹)	P^a	K^a (mg kg⁻¹)
Crop Failure on Sedge Peat Transitional Soil with Contact Layer at a Depth of 53 cm over Sand										
O_{1p}	68	0–13	3.8	0.5	78	32	0.30	29.1	9.5	25.9
O_2	42	13–53	3.5	3.9	90	31	0.26	10.9	0.5	19.6
O_3 (contact)	698	53–58	3.9	6.2	27	5	0.16	10.1	22.0	12.6
D_1	904	59–100	4.1	6.1	15	13	0.47	n/a	2.9	2.7
Crop Failure on Cotton Grass Peat Transitional Soil with Contact Layer at a Depth of 59 cm over Sand										
O_{0p}	101	0–3	3.8	3.5	124	23	0.16	11.7	3.1	13.8
O_{1p}	34	3–11	3.1	4.8	124	23	0.16	11.7	0.9	22.4
O_2	57	11–26	3.0	5.7	127	20	0.14	9.9	0.5	9.6
O_3	48	26–41	3.4	n/a	n/a	n/a	n/a	n/a	n/a	n/a
O_4	74	41–59	3.6	6.9	82	23	0.22	13.3	0.2	7.1
O_5 (contact)	116	59–65	3.6	6.0	38	4	0.10	n/a	0.3	10.5
O_6 (contact)	496	65–71	3.6	12.0	27	13	0.30	n/a	2.6	9.9
Cg_1	676	71–110	3.8	10.0	29	13	0.30	n/a	1.7	7.0
Cg_2	974	110–117	4.5	0.1	3	1	0.20	n/a	1.6	1.3
Cg_3	906	117–124	4.0	7.3	16	7	0.30	n/a	19.0	19.0
Successful Crop on Woody–Grassy Peat Transitional Soil over Sand										
O_{1p}	175	0–12	3.8	0.5	51	49	0.49	1.17	8.0	13.6
Og_2	655	13–25	4.0	3.6	61	33	0.35	0.48	3.5	11.1
Cg_1	993	25–64	4.6	0.0	2	2	0.44	n/a	1.6	2.0

Note: n/a = not available.
[a] HCl 0.2 M-extractable
[b] Exchangeable acidity
[c] Sum of cationic bases
[d] Cationic base saturation (cmol$_c$ kg⁻¹ per cmol$_c$ kg⁻¹)

Table 9.6 Grouping of Soils of Cutover Peatlands by the Degree of their Agricultural Improvement

Parameters	Assessment of Soil Quality			
	Poor	Medium	Good	Very good
cm				
Thickness of the arable layer	<15	15–20	20–30	30–40
Thickness of residual peat layer	<20	20–30	30–40	>40
Unitless				
C/N ratio	>25	18–25	14–18	10–14
pH Unit				
pH_{KCl}	<4.5	4.5–5.5	5.5–6.0	>6.0
g cm^{-3}				
Bulk density	<0.20	0.20–0.30	0.30–0.40	>0.40
g kg^{-1}				
Ash content (g/kg^{-1})	<100	100–200	200–400	>400
%				
Degree of peat decomposition (v/v)	<20	20–35	35–50	>50
Cationic base saturation (cmol$_c$/kg^{-1} per cmol$_c$/kg^{-1})	<0.50	0.50–0.60	0.60–0.75	>0.75
mg kg^{-1}				
0.2 M HCl-extractable Fe	>140	84–140	56–84	42–56
0.2 M HCl-extractable P	<9	9–15	15–26	>26
0.2 M HCl-extractable K	<25	25–42	42–58	>58

Source: From Kreshtapova, V.N. and Krupnov, R.A. 1998. Genetic peculiarities and basics of reclamation of cutover peatlands in Central Russia, in *Peatland Restoration and Reclamation: Proc. of the 1998 International Peat Symposium.* Malterer, T.J., Johnson, K., and Stewart, J., Eds., IPS Publ., Duluth, MN,115–119. With permission.

The pH (in KCl 1 M) providing optimal growing conditions for winter rye, oat, timothy, vetch, foxtail, lupine, potato, and Swedish turnip ranges from 4.8 to 7.0. Four years after reclaiming cutover peatlands near Tver, Russia, the cationic base saturation in the arable layer increased from 49 to 69% in sedge-*Hypnum* peat. Productive cutover peatlands show cationic base saturation greater than 60–75%.

Cutover peatlands overlying limnic materials do not have a contact horizon. The introduction of 10 to 30% sapropel into peat increased the content in exchangeable Ca and K, and in mineral N. Mixing sapropel with peat of pH >5.0 at rates of 5 to 20 Mg sapropel ha^{-1} (i.e., 80 to 250 Mg ha^{-1} on a dry-weight basis) markedly decreased acidity, and increased nutrient as well as ash contents and bulk density; it also enhances the quality of cutover peatland soils. The contact horizon should be destroyed by rototilling or plowing to improve the physical quality of the soil. Mixing the peat with mineral substratum is recommended to reduce wind erosion hazards and increase ash content, bulk density, and amounts of available K and P in the arable layer.

III. CONCLUSION

Adverse factors that limit the productivity of cutover peatlands are:

- An abrupt transition from waterlogging to water deficit due to an impermeable contact horizon between the peat and the mineral subsoil
- Abrupt temperature fluctuations during the growing period due to thick (>50 cm) peat layers
- Uneven thickness of residual peat layers
- Considerable spatial and temporal variations in crop yield and quality
- Nutrient imbalance

Agricultural use is recommended on lowland floodplain and near-terrace cutover peatlands near large cities. Elsewhere, cutover peatlands should be converted to natural landscape or afforested.

REFERENCES

Fatchikhina, O.E. 1960. Change in physical and chemical properties of peat quarry soil (in Russian), in *Proc. Central Peat-Bog Station,* 1:186–203, Moscow, Russia.

Korenova, T.S. 1982. Geomorphological peculiarities of soils of cutover peatlands of the Orekhovo-Zuevo area in the Moscow region (in Russian) in *Proc. Problems of Agricultural Utilisation of Peat Deposits: Collective volume of works of the Central Peat-Bog Station,* 6:31–35, Moscow, Russia.

Kreshtapova, V.N. 1996. Physical, chemical and geochemical properties of cutover peat areas cultivated in agriculture within the Moscow region in *Proc. 10th Int. Peat Cong.,* 2: 174–183, Bremen, Germany.

Kreshtapova, V.N. and Krupnov, R.A. 1998. Genetic peculiarities and basics of reclamation of cutover peatlands in Central Russia, in *Peatland Restoration and Reclamation: Proc. of the 1998 International Peat Symposium.* Malterer, T.J., Johnson, K., and Stewart, J., Eds., IPS Publ., Duluth, MN, 115–119.

Kreshtapova, V.N. 1993. *Trace Elements in Peat Soils and Peat Landscapes of the European Russia.* Rossel'khozizdat, Moscow.

Krupnov, R.A. 1978. Classification of soils of cutover peatlands according to the degree of their improvement (in Russian). *Torfyanaya promyshlennost',* 3:24–27.

Krupnov, R.A. and Popov, M.V. 1995. Reclamation of cutover peatlands (in Russian). *Bull. Tver State Technical University,* Tver, Russia, 3–80.

McNally, G. 1995. The utilisation of industrial cutaway peatland. The key factors influencing the various afteruse options, in *Proc. Conf. Peat Industry and Environment,* 84–86, Tallinn, Estonia.

Parent, L.E. and Caron, J. 1993. Physical properties of organic soils, in *Soil Sampling And Methods of Analysis. Can. Soil Sci. Soc. Methods Manual.* Carter, M.R., Ed., Lewis Publishers, Boca Raton, FL, 441–458.

Rin'kis, G. and Nollendorf, V. 1982. *Balanced Plant Nutrition by Macro- and Microelements* (in Russian). Zinatne, Riga, Latvia, 3–156.

Sapek, A. and Sapek, B. 1988. The methodological problems connected to chemical analysis of peat and peat soils, in *Proc. 8th Int. Peat Cong.,* 4:280–287, Leningrad.

Trutnev, A.G. 1963. *Cropping of Agricultural Plants on Cutover Peatlands* (in Russian). Sel'khozgiz., Moscow, 128 pp.

Vertogradskaya, T.A., Shirokikh, T.G., and Shirokikh, A.A. 1992. Ecological aspects of peatland soils and cutover bogs cultivation, in *Proc. 9th Int. Peat Cong.,* 1:215–222, Uppsala, Sweden.

CHAPTER 10

Agricultural Production Systems for Organic Soil Conservation

Piotr Ilnicki

CONTENTS

ABSTRACT

Largest areas of farmed organic soils in Europe are found in Russia (70,400 km²), Germany (12,000 km²), Belarus (9631 km²), Poland (7620 km²), and the Ukraine (5000 km²). In comparison, cultivated organic soils in United States and Canada cover 3080 km² altogether. In Europe, the agricultural use of organic soils takes 14% of total peatland area. Climatic factors limiting agricultural production on organic soils, a food production surplus and a serious environmental crisis led to European Union Directive No. 2078/92 intended to exclude large areas of peatlands from agricultural production. In most European countries, arable land use is advised only for shallow (< 1.0 m) or very shallow (< 0.5 m) peat deposits, or sand cover peat cultivation. Organic soil subsidence, a key factor in soil conservation, is primarily related to groundwater level. Depending on climatic conditions, intensity of drainage, peat type and land management, the annual loss of elevation is in the range of 0.3–1.0 cm yr^{-1} for grassland, and 1.0–5.0 cm yr^{-1} for arable land. Grassland is given priority in Europe due to shallower drainage, protection against frost, as well as reduced peat mineralization, CO_2 and NO_x emissions, and nitrate leaching.

I. INTRODUCTION

Peat is regarded as an important energy source in Finland, Ireland, Russia, and Sweden. In central and northern Europe, peat is also excavated for producing horticultural substrates for greenhouses and mushroom-growing cellars. In northern Europe, a large proportion of mires is covered by commercial forests. In Scotland, Highland heather peatlands are used as hunting grounds.

Organic soils are cultivated in northern countries such as England, southern Norway, Sweden and Finland, in the southern fringe of Karelia, and in central Russia up to the Moscow region; there, raised bogs dominate in the North, and fens in the South. Intensive agricultural use of peatlands is found in The Netherlands, Germany, Poland, Belarus, and the Ukraine. The capability of organic soils for agricultural production depends on climate, requirements for nature protection, mire geomorphology and vegetation, peat stratigraphy, stage of the moorsh forming process, the air-water regime, and soil physico-chemical properties.

A large food production surplus and a serious environmental crisis led to a European Union directive (Council Regulation No. 2078/92) to exclude peatlands from agricultural production. As a result, large areas of peatlands were converted to sylviculture and nature conservation considering high drainage costs, high nitrate content in some vegetables, and cereals lodging. In Poland, national parks were established on 66,000 ha in the Biebrza and Narew River Valleys. In Germany, a national park was created in the lower Odra valley together with a nature conservation area of 45,000 ha.

The aim of this chapter is to provide a European perspective for parsimonious use of peatlands in agriculture, considering biophysical and socioeconomic limitations.

II. AGRICULTURAL USE OF ORGANIC SOILS IN EUROPE

Macroclimatic factors limiting agricultural production on organic soils are:

1. Too short a vegetation period
2. Too low a mean annual temperature
3. Too large a difference between mean temperatures in July and January
4. Too large temperature differences between day and night

Table 10.1 Mean Annual Soil and Air Temperatures in the Biebrza River Valley and on Mineral Soils of Bialystok in the 1951–1965 Period

Location	Substrate	Height/Depth (cm)	Average Temperature °C January	July	Annual
Mineral soils in Bialystok	Air	+200	−5.1	17.5	6.4
	Soil	−5	−2.1	19.8	8.0
		−10	−1.9	19.6	8.0
		−20	−1.5	19.3	8.0
		−50	0.1	18.2	8.1
Organic soils in Biebrza	Air	+200	−4.9	16.5	6.0
	Soil	−5	−1.8	17.5	7.2
		−10	−1.3	17.1	7.1
		−20	−0.1	15.9	7.0
		−50	2.4	13.3	7.2

Source: From Kossowska-Cezak, U., Olszewski, K., and Przybylska, G. 1991. *Zeszyty Problemowe Postepow Nauk Rolniczych,* 372:119–160. With permission.

5. Frequent frost
6. Not enough accumulated degree-days during the vegetation period

The microclimate is also more severe in organic than in surrounding mineral soils. Organic soils are developed in lower topographical positions characterized by a higher temperature amplitude, higher frequency of frost, and higher relative air humidity. Thermal capacity is higher, and thermal conductivity is lower, in organic than in mineral soils. Thermic properties of the surface soil layer (0 to 0.2–0.3 m) depends on the volume occupied by organic and mineral substances, water, and air. Organic soils are cooler than mineral soils during the summer months and warmer during the winter (Table 10.1).

Compared with mineral soils, the range of agricultural production on organic soils is shifted to lower altitudes or latitudes. Land use data for organic soils are still scattered and frequently not reliable (Lappalainen, 1996). Areas of organic soils under agricultural use decreased in Europe due to economic reasons and to the need for nature protection (Table 10.2). The largest areas are found in Russia (70,400 km²), Germany (12,000 km²), Belarus (9631 km²), Poland (7620 km²), and the Ukraine (5000 km²). Cultivated organic soils in the United States and Canada cover 3080 km² altogether (Lucas, 1982). Agriculture occupies 14% of European peatlands, but 98% in Hungary, 90% in Greece, 85% in The Netherlands and Germany, and 70% in Denmark, Poland, and Switzerland. Meadows and pastures are considered to be the most effective conservation practice (see Table 10.3 for Poland). Arable lands are found mainly in Germany, The Netherlands, and Belarus.

III. PROFILE OF CULTIVATED ORGANIC SOILS

In Europe, toward the East from the Elbe River, only fens are in agricultural use. In countries with maritime climates, both fens and raised bogs are used for agricultural production. In The Netherlands and Germany, reclamation methods were

Table 10.2 Peatland Used for Agriculture in European Countries

No.	Country	Total Area (km²)	Peatland Area Used for Agriculture (km²)	(%)
1	Belarus	23,967	9631	40
2	Czech Republic and Slovakia	314	ca 100	ca 30
3	Denmark	1420	ca 1000	ca 70
4	Estonia	10,091	ca 1300	13
5	France	ca 1100	ca 660	ca 60
6	Finland	94,000	ca 2000	2
7	Germany	14,200	ca 12,000	85
8	Great Britain	17,549	720	4
9	Greece	986	ca 900	ca 90
10	Hungary	1000	975	98
11	Iceland	10,000	ca 1300	13
12	Ireland	11,757	896	8
13	Latvia	6691	ca 1000	15
14	Lithuania	4826	1900	39
15	Netherlands	2350	2000	85
16	Norway	23,700	1905	8
17	Poland	10,877	7620	70
18	Russia	568,000	70,400	12
19	Spain	383	23	6
20	Sweden	66,680	3000	5
21	Switzerland	224	ca 160	ca 70
22	Ukraine	10,081	ca 5000	ca 50
	Total	880,196	124,490	14.1

Note: ca = approximately.

Source: From Lappalainen, E., Ed. 1996. *Global Peat Resources.* Int. Peat Soc. Geological Survey of Jyskä, Finland, 359 pp. Changed and supplemented. With permission.

developed with partial or total reconstruction of the soil profile such as sand cover cultivation and deep ploughing.

The early Dutch fehn cultivation of raised bogs used in The Netherlands and Germany completely transformed the organic soil profile. A profitable alternative to peat burning in populated areas at the end of the 16th century, the Dutch fehn cultivation lasted till the beginning of the 20th century. Fibric peat was removed from mire surface, and the underneath sapric black peat was mined for fuel down to the subsoil. The top fibric peat used as filling material was applied onto the subsoil (40 cm thick). Sand and city wastes were added to make up an arable layer (10–14 cm thick) on top of the fibric peat, thus reconstituting an agricultural soil. At the end of the 19th century, the German peatland cultivation methods transformed bog and fen peat materials into productive soils with proper liming and fertilization; however, ploughing and mixing of the soil profile was required to improve root penetration due to low peat permeability (Figure 10.1).

A sand covering method was proposed in 1860 by Rimpau for the shallow fens of East Germany. The sand was excavated from bottom of a dense ditch network (15–20 m). Mechanized sand cover cultivation was later conducted in thick fens and

Table 10.3 Use of Fens (10,126 km²) and Bogs (751 km²) in Poland

Use	Fen (km²)	Fen (%)	Bog (km²)	Bog (%)
Forest	917	9.1	353	47.0
Grassland	7436	73.4	129	17.2
Arable land	55	0.5	—	—
Undrained peatland	1290	12.7	215	28.6
Cutover peatland	428	4.2	54	7.2

Source: From Lipka, K. 1984. *Studia Kom. Przest. Zagosp. Kraju*, 85:56–77. With permission.

bogs (up to 2.4 m thick) using the Rathjens Kuhlmaschine, a subsoil conveyor. The Rathjens machine dug down to 3.5 m in the soil to bring 1 m³ sand per m length to the surface, and spreading sand to form a sand layer about 10 cm thick. The German sand mix cultivation of raised bogs by deep ploughing resulted in alternate rows of peat and sand at an angle of 135° and in sand covering the peat. The peat:sand ratio was between 2:1 (coarse sand and fibric peat) and 1:2 (fine sand and sapric peat). Organic matter content at the surface was 6 to 8%. These methods improved the air–water regime, the microclimate, and soil carrying capacity in an area over 300,000 ha in The Netherlands and northwestern Germany (Emsland). In most European countries, arable land use is advised only for shallow (<1.0 m) or very shallow (<0.5 m) peat deposits, or sand cover cultivation. In Belarus, however, thick organic soils are often used as arable land, and shallow soils as grassland.

A soil profiling technique called "land crowning" was developed in Sweden for draining peatlands above the polar circle (Berglund, 1996). The shallow upper layer is ploughed in the direction of the centre to form a narrow bed (to 10 m wide). After a few years of soil modeling, the central part of the bed was elevated, and furrows could drain excess water on both sides of the bed.

Physicochemical properties of soil surface materials (0–30 cm) depend on peat type, pH and fertility, stage of the moorsh-forming process (MFP), and contamination. The MFP is slowest in soils under meadow and pasture, and for high water table levels (i.e., < 60 cm deep) (Okruszko, 1993). In Poland, most favorable conditions for agricultural production are obtained in fens with low to medium ash content (< 20%), fibric to hemic peat materials, low to medium MFP degrees, and slightly acid to neutral soil pH values (5.5–7.0 in 0.1 M CaCl$_2$). In some shallow organic soils, sediments containing significant amounts of sulphur or calcareous gyttja hinder root penetration.

IV. WATER BALANCE

A. Peatland Drainage

Mires can be classified according to water supply (rain, flowing water, spring water, groundwater), and groundwater fluctuations (Kulczynski, 1949; Moore and

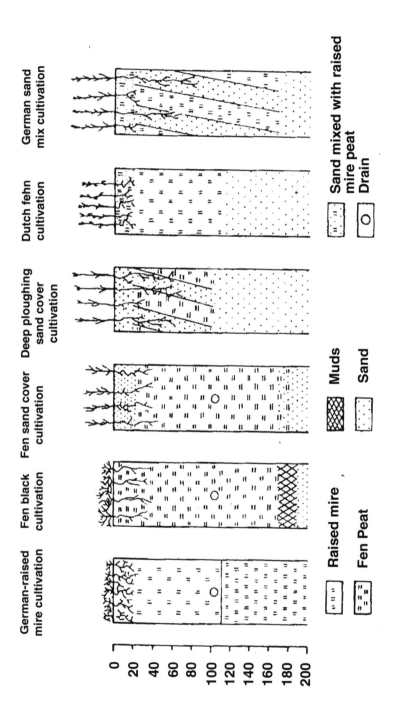

Figure 10.1 Root development in peatland profiles modified by cultivation methods. (From Göttlich, Kh., Ed. 1980. *Mire and Peat Science* (in German). E. Schweizerbartsche Verlagsbuchhandlung. Stuttgart, Germany. With permission.)

Bellamy, 1974; Göttlich, 1980; Dembek and Oswit, 1996). After mire drainage, soil phase distribution changes from around 5% solid volume and 95% pore volume (2% as air and 93% as water), to about 10% solid volume and 90% pore volume. To achieve 20% air and 70% water contents in organic soils for growing crops, 1840 m^3 water ha^{-1} may be evacuated for a drain depth of 80 cm (Göttlich, 1980). Due to lower permeability in organic compared with mineral soils in northwestern Germany, infiltration water is less than 30 mm yr^{-1} in raised bogs and 30–60 mm yr^{-1} in fens, compared with 100–200 mm yr^{-1} in sandy soils (Eggelsmann, 1973a). Because drainage and compaction may also reduce peat hydraulic conductivity, water partitioning between infiltration and runoff may further change.

Meadows and pastures require a water table drawdown to 0.4–0.8 m, while the water level could be as deep as 1.0–1.2 m for arable crops. By lowering the groundwater table 0.5–1.5 m below soil surface, soil water retained at suction less than pF 1.8–2.0 is evacuated through a network of ditches and canals. Ditch spacing depends on peat thickness and permeability. Until the end of the 19th century, draining consisted of narrow (0.1–0.2 m) and deep (0.8–1.0 m) slits cut through peat by hand. Later, mole drains were recommended. Slits and mole drains were replaced by more durable drains made of ceramic or plastic pipes. Drains could be wrapped with filtration materials to prevent silting (Eggelsmann, 1973b). Pumping systems were designed for small (up to 50 ha) or large (hundreds or thousands of ha) polder areas. Because small pumping facilities required a shallower network of main ditches and thus maintained more uniform water table levels across the area, they were more favorable to organic soil conservation compared with large systems.

B. Flooding and Runoff

Peatland drainage affects to some degree the water balance of catchment areas by increasing flood hazards. Compared with an undrained analog, a drained raised bog at Königsmoor, Germany, showed lower groundwater level and similar runoff during winter, but larger runoff at the end of the summer (Eggelsmann, 1990). In the raised bog of Chiemseemoor in Bavaria, the undrained portion discharged water at a slower rate than the drained analog (Schmeidl et al., 1970) (Table 10.4); runoff was higher in the drained portion of the bog during high rainfall and smaller during droughts.

For a mean annual precipitation of 729 mm during the 1968–1979 period in a drained raised bog at Ritschermoor in the Elbe River Valley, Germany, mean runoff was 341 mm, about 100 mm higher than the annual climatic water balance of +238 mm, due to an underground water supply (Ilnicki and Burghardt, 1981). In the case of a negative climatic water balance, annual runoff was about 250 mm, indicating a plateauing of summer flow. In undrained raised bogs, surface runoff may dominate because peat is saturated almost year round (Eggelsmann, 1990). During a dry spell, water discharge may stop. Five to 10 years after drainage, water balance resembled the original one (Eggelsmann, 1990). A high proportion of peatland areas in the catchment could decrease their water runoff during May–February and increase their maximum spring runoff (Ferda, 1973). Distribution of runoff values for the Biebrza River draining about 100,000 ha of peatlands in northeastern Poland indicated a

**Table 10.4 Change in Water Balance after Drainage of a Raised Bog at
Chiemseemoor, Germany, during the 1959–1968 Period**

Land Use	Water Runoff (mm)			Groundwater Level (cm)	PET[a] (mm yr⁻¹)
	Year	Winter	Summer		
Undrained with *Sphagnetum medii*	808	368	440	17.6	655
Drained — meadow	843	401	442	32.8	604

[a] Potential evapotranspiration.

Source: From Schmeidl, H., Schuch, M., and Wanke, R. 1970. *Schriften für Kuratorium Kulturbauwesen,* 19:1–171. With permission.

plateau during summer months. The plateau was higher the larger the catchment area, more distinctly in dry than in wet years (Byczkowski and Kicinski, 1991).

V. SUBSIDENCE

A. Initial Subsidence

Subsidence of organic soils first results from loss of buoyancy upon drainage. Successive soil drying and wetting cycles cause irreversible peat shrinkage and swelling leading to fissures and a granular structure. Peat materials change into *Mursz* (in Polish), *vererdete-Torfböden* (in German), *Terre noire* (in French), and muck or earthy peat (in English). The term *moorsh* was proposed by Henryk Okruszko (Okruszko, 1993) to describe the material derived from MFP.

According to Segeberg (1962), peat mineralization proceeds until ash content of surface soils reaches 900 g kg⁻¹, the target for sand cover cultivation. Subsidence is slower in deeper peat layers. There are two phases of peatland subsidence after drainage as follows:

1. Initial subsidence is caused by load and shrinkage of upper peat layers depending on drainage depth (Segeberg, 1960).
2. Microbial oxidation consumes organic soils in the long run as influenced by soil type, temperature, and groundwater level (Mundel, 1976).

Due to uneven moisture distribution, subsidence is not spatially uniform. Subsidence is intensified by peat fires, wind and water erosion. Peatland subsidence rate is higher the thicker the peat strata, the lower the peat bulk density, and the deeper the free draining ditches.

Soil subsidence during the first phase (5 to 10 years after drainage) is frequently calculated by one of the three following empirical equations obtained by measuring subsidence and its main causal factors in drained peatlands:

Hallakorpi equation (Finland): $S = a(0.080T + 0.066)$ (10.1)

Panadiadi-Ostromecki equation (Eastern Europe): $S = \sqrt[3]{T A d_{gw}^2}$ (10.2)

Table 10.5 Constants Related to Peat Bulk Density in Subsidence Models: S as Subsidence (m), T as Peatland Thickness before Drainage (m), d as Ditch Depth after Drainage (m), and d_n is Drainage Depth after Subsidence (m)

Hallakorpi Model		Panadiadi–Ostromecki Model		Segeberg Model	
Bulk Density		Bulk Density		Solid Volume	
Qualitative	a^a	(g cm^{-3})	A^b	(% v/v)	k^c
Loose	2.85	0.079	0.49	<5.0	0.303
Rather loose	2.00	0.093	0.35	5.0–7.4	0.219
Rather dense	1.40	0.109	0.25	7.5–12.0	0.154
Dense	1.00	0.128	0.18	>12	0.113

[a] $S = a(0.080T + 0.066)$

[b] $S = \sqrt[3]{TAD^2_{gw}}$

[c] $S = k\, d_n T^{0.707}$

Source: From Ilnicki, *Wiadomouci Mel.*, 1965, vii:57–61. With permission.

Segeberg equation (Germany, The Netherlands): $S = kdT^{0.707}$ (10.3)

where S is subsidence (m); a, A, k are constants depending on peat density (Table 10.5); T is peat thickness before drainage (m); the Panadiadi-Ostromecki d_{gw} is the lowering of groundwater (here ditch depth) after drainage (m); the Segeberg d is drainage depth after subsidence (m), computed by difference between groundwater levels before and after drainage. Hallakorpi's formula assumes a drainage depth of 1.1 m. Equations 10.1 to 10.3 were elaborated from varied parameters in recently drained peatlands, but could be used for a second drainage phase.

B. Long-Term Subsidence

The second stage is dominated by the biologically driven peat decay. The intensity of the long-term subsidence depends on drainage intensity, climatic conditions, land use and management, peat bulk density, and, to some extent, the botanical composition influencing MFP.

Ilnicki (1977) found that power equations for organic soil subsidence for 30 years (S in cm) at the Ritschermoor mire depended on bulk density as follows:

Loose peat materials: $S = 14.3t^{0.442}$ (10.4)

Rather loose peat materials: $S = 7.35t^{0.478}$ (10.5)

Rather dense peat materials: $S = 5.14t^{0.485}$ (10.6)

Long-term studies conducted in the Notec Valley led to models describing the subsidence of deep organic soils made of reed peat and used as meadows (Ilnicki, 1973). During the 1903–1969 period, subsidence rate was 0.33 cm per year with shallow ditches (0.4–0.6 m) and 1.12 cm yr^{-1} with deep ditches (1.0–1.2 m).

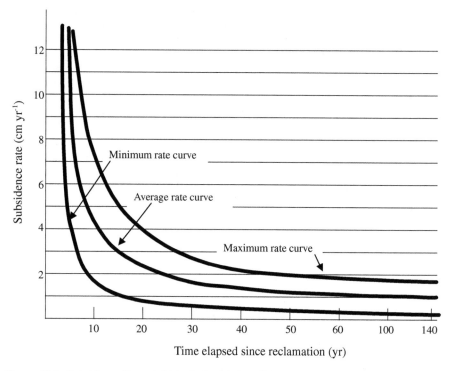

Figure 10.2 Subsidence through biological oxidation. (Adapted from Maslov, B.S., Kolganov, A.V., and Kreshtapova, V.N. 1996. *Peat Soils and Their Change under Amelioration* (in Russian). Rossel'khozizdat Ed., Moscow, 147 pp. With permission.)

Schothorst (1977) studied peatland pulsation in deep fens drained between the 9th and 14th centuries in The Netherlands, and now used as pastures. By comparing bulk density of organic matter in layers above and below groundwater level, he estimated that 15% of total subsidence of 2 m over the past 1000 years could be ascribed to shrinkage of the upper layer, and 85% to oxidation of organic matter. The rate of organic matter loss was 2 mm yr^{-1} for high water table (0.2–0.5 m), up to 6 mm yr^{-1} with deeper drainage.

Stephens (1960) and Harris et al. (1961) showed that subsidence resulting from decomposition of organic matter was related to groundwater level. Peat mineralization in German fens was investigated by Mundel (1976). The greatest influence was exerted by the water table level and soil temperature. Highest rate of organic matter decomposition (490 g C m^{-2} yr^{-1} or 0.3 cm yr^{-1}) was associated with a 90-cm groundwater level. In Florida and California, subsidence rate under vegetable cropping reached 7 cm yr^{-1} for a 1.0 m-deep groundwater level (Stephens, 1960). In Israel, agricultural use of peatland can lead to an oxidation rate up to 10 cm yr^{-1} (Levin and Shoham, 1972). Subsidence rate of drained organic soils is typically found in the range of 0.3–1.0 cm yr^{-1} for grassland, and 1.0–5.0 cm yr^{-1} for arable land. A synthesis of subsidence data of the second phase (microbial decomposition) is shown in Fig. 10.2 (Maslov et al., 1996). The values range from 1 to 7 cm yr.$^{-1}$

Maximum values were obtained for peat with low ash content and bulk density of 0.08–0.10 g cm^{-3}, warm climate, and arable farming.

VI. BEST MANAGEMENT PRACTICES FOR ORGANIC SOIL CONSERVATION

Optimum use of organic soils must take into account several natural, social, and economic considerations. Natural conditions are temperature, air–water regime, plant communities, peat stratigraphy, soil properties, as well as valuable species and ecosystems. National security, population density, demand for food, and agricultural policy are the social conditions. The economy of peatland use includes cost for fuel and infrastructure, competing uses, production efficiency, biodiversity and economic policy.

Arable farming is expensive to establish on vast peatland areas requiring the construction of infrastructures. Meadows and pastures need shallower drainage, are more resistant to frost, and less subject to peat mineralization, emission of carbon dioxide and nitric oxides, and nitrate leaching.

In Europe, the range of field, horticultural and nursery crops grown on organic soils is usually limited to potatoes and spring crops. In North America, crop range is much larger (Lucas, 1982). Vegetables grown in organic soils may accumulate nitrates in their tissues. Advantages of growing vegetable crops in organic soils include large water supply and low-energy requirement for tillage. In western Europe, arable farming on organic soils is facilitated by sand cover, which increases soil bearing capacity. The largest problem with field crop agroecosystems is the accelerated mineralization of organic matter leading to fast decession of organic soils.

Meadows and pastures occupy 80 to 85% of total peat farmland in Europe. They do not require deep drainage, thus slowing down MFP and conserving soil. In the long run, however, the meadow also leads to peat wastage. Grassland covers all types of organic soils. In England, The Netherlands, and Germany, pasture imposes a permanent pressure from cloven hooves of animals, limiting peat mineralization even more than in meadows. In eastern Europe, meadows were established because of a more severe climate on organic than on mineral soils, and of the impossibility to pasture animals on remote organic soil areas. Compared with arable farming, grassland increases biodiversity and supplies refuge to waterfowl and rare mammals. In other words, when agricultural use of peatland is considered in Europe, grassland is given priority.

VII. CONCLUSION

The capability or organic soils for agricultural production depends on many natural and socioeconomic conditions. Nowadays, agriculture uses only 14% of European peatlands. Depending on climatic conditions, intensity of draining, peat type and land management, the annual loss of elevation is in the range of 0.3–1.0 cm yr^{-1} for grassland, and 1.0–5.0 cm yr^{-1} for arable land. The wisest use of organic soils in Europe is grassland or wetland restoration.

REFERENCES

Berglund, K., 1996. Peatland drainage for agricultural purposes in the north of Sweden. *Proc.10th Int. Peat Congr.* 1:79.

Byczkowski, A. and Kicinski, T. 1991. The hydrology and hydrography of the Biebrza River basin (in Polish). *Zeszyty Problemowe Postepow Nauk Rolniczych,* 372:75–118.

Council Regulation 2078/92. 1993. Agricultural production methods compatible with the requirements of the protection of the environment and the maintenance of the countryside. *Official J. European Commun.,* L 215/85.

Dembek, W. and Oswit, J. 1996. Hydrological feeding of Poland's mires. *Proc. 10th Int. Peat Congr.,* 2:1–12.

Eggelsmann, R. 1973b. The thermal constant of different high-bogs and sandy soils. *Proc. 4th Int. Peat Congr.,* 3:371–382.

Eggelsmann, R. 1973a. *Drainage Instructions. Agriculture, Engineering, Landscape Management* (in German). Verlag Wasser und Boden, Axel Lindo, Hamburg, Germany.

Eggelsmann, R., 1990. Water regulation in mires (in German), in *Moor- und Torfkunde.* Göttlich, Kh., Ed., E. Schweizerbartsche Verlagsbuchhandlung. Stuttgart, Germany, 321–349.

Ferda, J. 1973. Hydrological function of mires in mountainous areas (in German). *Zeitschrift für Kulturtechnik und Flurbereinigung,* 14:178–189.

Göttlich, Kh., Ed. 1980. *Mire and Peat Science* (in German). E. Schweizerbartsche Verlagsbuchhandlung. Stuttgart, Germany.

Harris, C.J. et al. 1961. Water level control in organic soils, as related to subsidence rate, crop yield and response to nitrogen. *Soil Sci.,* 94:158–161.

Ilnicki, P. 1965. Peatland subsidence. (in Polish).*Wiadomosci Mel.* Łak.3.VII:57–61.

Ilnicki, P. 1973. Subsidence rate of reclaimed peatlands in the Notec River Valley (in Polish with English summary). *Zeszyty Problemowe Postepów Nauk Rolniczych,* 146:33–61.

Ilnicki, P. 1977. Subsidence in repeatedly drained bog mires of northwestern German flatlands, 3. Report: Assessment (in German). *Zeitschrift für Kulturtechnik und Flurbereinigung,* 18:153–165.

Ilnicki, P. and Burghardt, W. 1981. Subsidence in repeatedly drained bog mires of northwestern German flatlands, 6. Report: Influence of climatic water budget and flow on relief formation, and on surface and drain subsidence (in German). *Zeitschrift für Kulturtechnik und Flurbereinigung,* 22:112–121.

Kossowska-Cezak, U., Olszewski, K., and Przybylska, G. 1991. Climate in the Biebrza Valley (in Polish). *Zeszyty Problemowe Postepow Nauk Rolniczych,* 372:119–160.

Kulczynski, S. 1949. Peat bogs of Polesie. *Mémoires de l'Académie Polonaise des Sciences et des Lettres,* Série B:1–356.

Lappalainen, E., Ed. 1996. *Global Peat Resources. Int. Peat Soc.* Geological Survey of Jyskä, Finland, 359 pp.

Levin, I. and Shoham, D. 1972. Nitrate formation in peat soils of the reclaimed Hula Swamp in Israel. *Proc. 4th Int. Peat Congr.,* III:47–57.

Lipka, K. 1984. Economic opinion about peat deposits in Poland (in Polish). *Studia Kom. Przest. Zagosp. Kraju,* 85:56–77.

Lucas, R.E. 1982. Organic soils (histosols). Formation, distribution, physical and chemical properties and management for crop production. Research Report 435, Michigan State University, East Lansing, MI, 77 pp.

Maslov, B.S., Kolganov, A.V., and Kreshtapova, V.N. 1996. *Peat Soils and Their Change under Amelioration* (in English). Rossel'khozizdat Ed., Moscow, 147 pp.

Moore, P.D. and Bellamy, D.J. 1974. *Peatlands.* Elek. Sci., London.

Mundel, G., 1976. Mineralization of fen peat (in German). *Archiv für Acker Pflanzenbau und Bodenkunde,* 20:669–679.

Okruszko, H. 1993. Transformation of fen-peat soil under the impact of draining. *Pol. Akad. Nauk,* 406:3–75.

Schmeidl, H., Schuch, M., and Wanke, R. 1970. Water content and climate of cultivated and pristine bog mires in the Alps (in German). *Schriften für Kuratorium Kulturbauwesen,* 19:1–171.

Schothorst, C.J. 1977. Subsidence of low moor peat soils in the western Netherlands. *Geoderma,* 17:265–191.

Segeberg, H. 1960. Peatland subsidence after drainage and assessment with empirical equations (in German). *Zeitschrift für Kulturtechnik und Flurbereinigung,* 1:144–161.

Segeberg, H. 1962. Assessment of subsidence by peat shrinkage (in German). *Zeitschrift für Kulturtechnik und Flurbereinigung,* 3:356–367.

Stephens, J.C. 1960. Subsidence of organic soils in the Florida Everglades. *Soil Sci. Soc. Am. Proc.,* 24:77–80.

Index

Milton Keynes UK
Ingram Content Group UK Ltd.
UKHW040059071024
449327UK00019B/673